German Annual of Spatial Research and Policy

Editors:

ARL – Academy of Regional Sciences and Regional Planning, Hanover
BBR – Federal Office for Building and Regional Planning, Bonn
IfL – Leibniz Institute for Regional Geography (IfL), Leipzig
IOER – Leibniz Institute of Ecological and Regional Development, Dresden
IRS – Leibniz Institute for Regional Development and Structural Planning, Erkner (near Berlin)

Responsible for the 2008 edition:

BBR – Federal Office for Building and Regional Planning

Wendelin Strubelt (Ed.)

Guiding Principles for Spatial Development in Germany

With 44 Figures and 6 Tables

Springer

Editor
Prof. Dr. Wendelin Strubelt
Federal Office for Building
and Regional Planning (BBR)
Deichmanns Aue 31-37
53179 Bonn, Germany
Fasanenstr. 87
10623 Berlin, Germany

Editorial staff
ts redaktion, Berlin: Thies Schröder, Sandra Kalcher

English language editor
Mark Kammerbauer, Berlin

Layout
Karin Engelke, Berlin

ISSN 1862-5738 e-ISSN 1862-572X
ISBN 978-3-540-88838-3 e-ISBN 978-3-540-88839-0
DOI 10.1007/978-3-540-88839-0
Springer Dordrecht Heidelberg London New York

Library of Congress Control Number: 2009921820

Cover design: eStudio Calamar S.L.

Printed on acid-free paper

Springer is part of Springer Science+Business Media (springer.com)

Acknowledgement

Upon completion of this book, I wish to extend my gratitude to the contributing authors, Jerzy Bański, Annedore Bergfeld, Clemens Deilmann, Marek Degórski, Hans-Jörg Domhardt, Ludger Gailing, Günter Herfert, Rupert Kawka, Tomasz Komornicki, Thomas Kuder, Gerd Lintz, Christoph Scheck, Karl Peter Schön, Przemysław Śleszyński, Gabi Troeger-Weiß, Grzegorz Węcławowicz, Marek Więckowski, Peter Wirth, and Horst Zimmermann, for their contributions to this publication. I would also like to thank Heiderose Kilper, Sebastian Lentz, Bernhard Müller and Dietmar Scholich, who have also supported this publication as editors. Further special thanks go to Sieglinde Lehmler and Beatrix Thul as well as to Sandra Kalcher and Mark Kammerbauer from the ts redaktion.

Wendelin Strubelt

Contents

1 Preface – Introduction

Wendelin Strubelt

For the third time, the five leading German research institutes for spatial planning and policy – beyond German universities – offer an international audience a focused overview over their engagement in scientific analyses and political counselling.

This time we have concentrated our papers around the topic of the development and the formulation of guiding principles, concepts, or visions – in German we use a term not yet well known in international contexts, *Leitbilder* – for spatial planning and policy. For the non-German audience it might be important to know that, within the German context, we are dealing with a rather complex federal system.

On the one hand, this complexity results in a very fragmented structure of formal and legal duties and actual decision-making beyond any established hierarchical assumptions for administrative efficiency. On the other hand, this form of a cooperating and competing political system with its different constituencies and bureaucracies allows for or supports different views to be developed and presented. Faced with such a fragmented national structure, the analysis of spatial planning and policy needs to carefully differentiate between the various levels of decision-making and the interactions between them. With the contributions at hand, we have assumed to cover this complex picture.

In this regard, this book reflects the use of concepts or visions – *Leitbilder* – for spatial planning and policy at the different federal levels. This includes the European level or cross-border relations at the macro level and the micro level of the local community as well. It includes rather theoretical reflections about the "philosophies" behind such concepts as well as the very concrete analysis of local concepts required by local communities with different historical traditions and backgrounds. It includes analyses about the physical background of urban planning and the concepts providing means of cooperation needed for spatial planning and policy along national borders.

Thus, the reader may find a lot of information and analyses from a very complicated field of policy-making, of conceptual framing of spatial development. Perhaps this picture of federal interacting seems strange to foreign readers, but it may provide valuable insight, especially if we take into consideration that such complicated interaction will comprise our future. On account of the necessary interactions in a Europe of regions, this implies the absence of a strict or single national and supra-national authority. The fading away of strict structures and the rise of fragmented structures throughout might very well be our future.

This book is a common endeavour of the five German institutes for spatial planning and policy. Again, it intends to strengthen the position of German spatial analysis in the international contexts. It is in the hand of our readers whether we succeed.

2 What is a „Leitbild“? Some Reflections on the Origin and Use of the German Expression

Horst Zimmermann

2.1 Purpose and Structure of the Paper

The German government document of 2006 on “Leitbilder und Handlungsstrategien in Deutschland“ was the result of two years of professional and political discussion (BMVBS 2006a: 1)[1]. The author was part of that process and, having analyzed much earlier processes of forming similar policy documents on spatial planning (Zimmermann 1966), felt stimulated to document the differences and similarities in dealing with such concepts.

The central concept is that of „Leitbild“ (plural “Leitbilder”), which dominates title and substance of the document of 2006. The paper therefore tries to lay open the roots of that concept first (B), and in particular to point out its both empirical and normative elements. The example of the new growth Leitbild of 2006 is then used to show how scientific arguments and political intentions can be kept separate in a fruitful new way (C). From this starting point, the new Leitbild generation of 2006 is then analyzed as to its content and function (D). The paper ends with a critical note on the use of the Leitbild notion as such (E).

The field of visions and concepts in spatial planning is very wide, and to fill it completely could easily lead to an entire book. Therefore, it is necessary to point out what the author did not intend to do[2]. First of all, no international comparison of the use of such concepts in practical spatial planning was meant to be carried out. The specialist will easily be able to point out that in some country and at some point in time the applied concept bears quite some similarity to a Leitbild. Second, it was not planned to embed the German development in the wider planning theory literature. Instead, the intention is (1) to lay open the roots of the somewhat specific German expression of “Leitbild” (with the somewhat special nature already to be gathered from the difficulty of translating the term adequately into English), and

(2) to demonstrate how, by continuous use of the term, a modern content has been given to it, helpful to the German discussion, exemplified by the Leitbild “Growth and innovation”. This narrow perspective was chosen due to the fact that the entire

---

1 For details of this document see the other contributions in this volume. – All literal quotations from German sources are translations by the author.

2 The author is grateful to an unknown referee who pointed out that in several instances the earlier version could have led to a wider expectation than the author had intended to meet.

volume happens to be dedicated to "Leitbilder", so the foreign reader might be interested in the origin and development of this expression.

2.2 A Term and its Historical Role

The roots of the Expression "Leitbild"

For a foreign reader with some knowledge of the German language, the key term „Leitbild" in relation to the spatial dimension, as it is used in the influential government document of 2006, must be puzzling. It does not correspond directly to any English equivalent[3]. It is also not very common and definitely not very old in the German language. The authoritative German dictionary Brockhaus circumscribes Leitbilder exclusively as rather personal ideas of man (Brockhaus 1990: 249) and not as political items.

At least in its spatial relation, the origin of the use of this term can be traced down precisely. E. Dittrich asserts that he first used the term in a lecture in 1953 (Dittrich 1953; according to Dittrich 1962: 2; see also Storbeck 1982: 212)[4]. He circumscribes Leitbilder as „mental formative principles, which bring all the social context elements together for a given historical epoch " (Dittrich 1962: 6; SARO 1961: 51). What is meant becomes clearer when he refers to the more general and non-spatial Leitbild of the social market economy, as it was introduced by A. Müller-Armack (Dittrich 1962: 16; SARO 1961: 67) and which was very influential in shaping Germany after World War II (see below).

To develop such a very general concept and to position it high up in a politically relevant setting has to be understood in connection with the historical background of the 1950s and early 1960s. Politically, one wanted to move away from the Nazi ideology, in this case from „the postulates of a holistic spatial research and spatial order" (Wurzer 1962: 4). Therefore it probably seemed helpful to refer to an overarching concept which contrasted with the concepts of the previous era[5]. But at the same time this discussion occurred in a particular historic phase and under given methodological perceptions. The notion of an „order" as some guiding principle was typical for this time, not least in the form of the „social market economy order" as mentioned before. To develop the notion of such an „order" involved strong value judgments. The discussion is therefore not purely scientific, but is heavily interwoven

3 This is the more astonishing as the similar term "Leitmotiv" is directly used in English (This aspect I owe to Manfred Sinz).

4 However, Dehne says that „presumably" the term Leitbild has already been used in the early 19. century, but he does not quote here (Dehne 2005: 608).

5 Dittrich himself had held a non-influential post in the academic planner scene during the Nazi period. But he was then one of the few who turned away immediately, and he was apparently the only one who admitted, for instance, that there had been a racist motive behind the settlement policy of the Nazis in the East (Leendertz 2008: 15-17).

with political statements or at least implicit political assumptions of the authors.

This does not provide problems as long as the discussion occurs between scientists and politicians, with both sides adding identifiable elements into an amalgamated product. What is disturbing in this context is the scientific interpretation of the way in which a Leitbild is derived. In the quoted sources, Dittrich clearly has the feeling that he, as a scientist, can develop such a Leitbild, not noticing that his personal value judgments enter the outcome and determine the result decisively. The methodological problem is: Ten other scientists would have developed ten other forms of Leitbild with respect to spatial conditions in Germany at that time. The normative content becomes evident when he says (in a twisted way, also in German): „If confronted with a task, one must be able to say what one wants, and not only, what one could want privately (without bearing special responsibility), but what one, when being responsible, must desire, or ought to desire, as being ‚in order' " (Dittrich 1962: 3). What clearly emerges from this reasoning is the message that the scientist is supposed to develop and introduce value judgments in the statements he gives on political issues.

All of this may sound artificial to a foreigner, but it is necessary to understand how science was dealing with political issues at that time. A scientist then often believed that in his function as a scientist he was able to make better value judgments than others, a notion which, in the Popper tradition of science, has been given up long since. And it remains to be seen how the new approach of 2006 avoids – or does not avoid – this pitfall.

The early role of the Leitbild in Spatial Planning

From what was said before, it is evident that a Leitbild is a concept which is positioned high up in the goal-instrument-hierarchy. Very influential in developing this hierarchy was a document from 1961, both powerful and, for spatial planning, influential - the report of the Expert Committee for Spatial Order (SARO 1961), commissioned by the federal government. In its terms of reference, the government explicitly demanded the Committee to specify a Leitbild, as all future government activities with spatial implications would have to be oriented towards it (SARO 1961: 7). As to the aforementioned separation of science and political intention, it is noteworthy that the SARO experts were not only scientists, but half of the ten members had been delegated from various ministries. The SARO report set out by defining two tiers of Leitbild, plus two additional steps.

The societal Leitbild

The Committee defined as top of the hierarchy a societal Leitbild with three elements: freedom, social balance, and security. They correspond closely to what had been formulated for the social market economy by Walter Eucken (Eucken 1952/1990). He had developed his famous set of concrete principles for a market economy against the background of a separate chapter on "Freedom and order" (Zimmermann 1994: 138). According to the SARO Committee, one element of this societal Leitbild is the Leitbild of Spatial Order.

The Leitbild of Spatial Order (Raumordnung)

Raumordnung is one of the most intensively discussed concepts in German spatial planning philosophy, especially if contrasted with Raumplanung. The latter is easy to translate as spatial planning. But what is in comparison meant by Raumordnung? Dittrich, whose ideas clearly entered the SARO report, fiercely argues that „spatial order" (Raumordnung) must never be confounded or mixed with „spatial planning" (Raumplanung), not to speak of such technical matters as planning instruments (Dittrich 1962: 3). Unfortunately it is very difficult to translate „Raumordnung" in a meaningful way. A very helpful volume on German-English planning vocabulary (Turowski 2002: 103) translates it in the same way as Raumplanung, namely by "spatial planning". Consequently, this meaning, which Dittrich attributed to Raumordnung in explicit confrontation to Raumplanung, is supposed to convey something more elaborate. Raumordnung is definitely not implying the empirical setting in a given spatial situation, as the geometrical understanding of the word „Ordnung" could suggest. It is a more philosophical notion, related to a Leitbild as the guiding principle, which in turn fills „Ordnung" with its non-empirical content[6]. This is similar to the use of the word "Ordnung" in relation to the social market economy as the central concept of the German "Ordnungstheorie", for which again there exists no adequate English translation. The term "Leitbild" itself is, with respect to spatial planning, translated as "perspectives" (Turowski 2002: 101) in the German-English planning vocabulary volume, which is probably as close as one can get to its meaning, though the German word "Perspektiven" also exists and is free of philosophical connotation, being closer to outlook or – well – perspectives in the usual English meaning[7].

The Principles of Raumordnung

From the Leitbild of spatial planning so-called principles of Raumordnung have been derived at a later stage. In the present federal law 15 such principles have been laid down. All fields of spatial activities are to be "developed", "secured", "balanced" etc. Looking at them more closely, these finally are tangible fields of policy with – still rather general – directions into which policies are expected to move. In SARO (1961: 54-63) the respective 14 items are called the details of the "Leitbild der Raumordnung."

6 Recent use of the term Raumordnung is more down-to-earth. Sinz (2005: 863) sees it as the position where "development perspectives beyond individual places and policy fields" are laid down.

7 For the recent use of the term "perspectives" see below section D I. – A translation of "Leitbild" as "vision" does not seem focused enough on intended activities. The Webster dictionary (Webster 1989: 1597) interprets the nearest meaning of "vision" as "the act or power of anticipating that which will or may come to be: prophetic vision; the vision of an entrepreneur".

The Level of Plans and Instruments

Leitbilder and principles then are intended to guide the development of the concrete planning procedures at the federal, *Länder* and local levels and the execution of such plans. Whether the respective plans then are already called instruments, or whether the instruments implement the plans, is a matter of further definition.

This hierarchical structure of the argument is by no means only of historical importance. The presently valid federal law in this field is called „Raumordnungsgesetz des Bundes" of 1997 and was last amended in 2004. It contains (1) a so-called Leitvorstellung, which clearly corresponds to the word Leitbild, and (2) principles of Raumordnung, which "are to be applied in the spirit of the Leitvorstellung" (§ 2 (1)). The Leitvorstellung itself refers to the three-pillar version of sustainability, which is a modern type of high-ranking goal and thus means some progress in rendering the concept more tangible. – What is interesting in this context is the fact that in § 18 of this law the federal government is supposed to develop "Leitbilder", and with the document mentioned in the beginning that is exactly what it did.

In historical perspective it is interesting that the Leitbild (or Leitvorstellung as the synonym) was not introduced in the first federal law of 1965, though the content of the pertinent § 1 was roughly the same as that of the next law of 1989. It was only in the latter that the term Leitvorstellung appears in § 1, which means that the Leitbild approach had even gained strength in the meantime.

2.3 The Recent Connection of the Leitbild Discussion with Theory: The Example of the "Growth and Innovation" Leitbild

A new Type of Discussion Process

As other contributions in this volume spell out more precisely, the federal government together with the *Länder* started a process which led to the Leitbild document of 2006. In the following it will be pointed out how the new document deals with the challenges which became visible from the past example: How has the scientist – or more precisely the theory of the time – been integrated in the process? How has the problem of value judgments been treated in this context? Such questions will be dealt with by looking at the example of the new growth Leitbild.

The basic document mentions two years of scientific and political discussion, together with analyses by the Federal Office for Building and Regional Planning (BBR) in Bonn (BMVBS 2006a: 1). At the heart of the process was Manfred Sinz in the Federal Ministry of Transport, Building and Urban Development (BMVBS). At all stages of this process he summed up the latest stage of the process, brought the intermediate results to the professional public and monitored the discussion process through several workshops, internal discussions etc. (see for instance Sinz 2006 and Aring and Sinz 2006).

To turn to the "Growth and innovation" Leitbild as one example of the new approach, this process started with the geographical presentation of past trends and driving forces, drawn and presented by the BBR in an analytical empirically-based way. In the subsequent workshops it became evident that the underlying regional economics had substantially changed over the past years.

From Convergence Theory to new Economic Geography

The theoretical framework for the previous decades of spatial planning had been the convergence theory[8]. Its essence can be illustrated in a metaphor. The agglomerations can be visualized as spilling over, so to speak. Labour and land are scarce and therefore expensive. The opposite is true for structurally weak peripheral regions, because these factors of production are available in overabundance there and thus are very reasonable in price. Consequently, these factors rush towards the peripheral regions. The economic result is first a narrowing of factor proportions and then of factor prices, both "converging" to the same levels. For spatial planning, this theoretical perception meant a great advantage, because the objectives of regional equalization and of furthering national growth were in harmony. If the poor regions were supported by regional policy, this was, from the perspective of this theory, at the same time good for national growth, because those regions were supported where growth tended to move anyway[9].

Since the 1990s a new perspective developed in regional economics. Paul Krugman and his New Economic Geography are quoted most often, but Grossman/Helpman should also be mentioned (Zimmermann 2004). In this view, an agglomeration is no longer seen as an accident during economic growth processes, to be amended soon by convergence. Instead, agglomerations are natural outcomes of the growth process, so to speak. The central argument is decreasing cost of transportation, for goods and services, and above all for information. This lower transportation cost allows many production processes to now occur in one place, whereas before they had to be spread over the country in order to lower the transportation cost. In addition, the regional concentration allows profiting from economies of scale. This in turn permits higher salaries for high-quality workforce, thereby further improving effectiveness and efficiency in agglomerations. Results from other disciplines, like the sociological research in "creative milieus" or the scientific results for the knowledge economy, have corroborated the basic hypothesis. This states that it is the modern agglomeration (defined spaciously) which is at the heart of future growth chances in high-income societies, and this can happen even if this agglomeration itself does not grow at a given point in time. Though growth also occurs in other regions and though the agglomeration process cannot go on ad

8 For the following see Zimmermann (2004).

9 The convergence theory has been said to still hold between EU member countries, but not inside the larger member countries (Puga 2001), for which Germany can serve as example.

infinitum, there seems to be widespread agreement on the great importance of these regions for growth.

For spatial planning, this result is disillusioning. Helping the poor regions is no longer simultaneously policy for national growth. Instead, the money has to be earned in the agglomerations (and other growth regions) first, which therefore have to be observed and supported in spatial planning. Only then can help for the poor regions be provided, and on purely distributional grounds.

From Analysis to Political Statement

These scientific results were entered into the discussion process for the new Leitbilder. They showed that sticking with the previous "equalisation-first"-philosophy would be very costly for the country in the long run.

So far it was science. In the discussion process the political side responded by developing schemes which indicated a way to further national growth. One of them was the definition of "European Metropolitan Regions" in Germany. As much as can be said against their large number and the realistic potential of some of them, developing the scheme certainly was one way of demonstrating that the growth objective was taken more seriously than before.

But what was more important and new was the fact that, based on the fact finding of BBR and the results of the theoretical discussion, the political side drew up a normatively complemented Leitbild-graph. It constituted what policy wanted to achieve in the long run, accepting some deficiencies, mainly in eastern Germany, and focusing on agglomerations and on the additionally defined growth centres outside metropolitan regions (sorting them out was an achievement of BBR). On the whole, the example of the growth Leitbild of 2006 demonstrates how the scientific content can be kept separate from the political statement, with the one permitting a sound format of the other.

2.4 2006: A Modern Version of a Leitbild

The Documents

The result of the Leitbild discussion in the early years of the millennium were various documents of BMVBS, prepared and sometimes (co-)authored by BBR. The decisive document was that of June 30, 2006 (BMVBS 2006a), adopted that day by the Standing Conference of Federal and State Ministers Responsible for Spatial Planning (MKRO). Its heading was "Leitbilder und Handlungsstrategien für die Raumentwicklung in Deutschland". A somewhat popularized version was published the same year in German and English under the – somewhat different – title "Perspektiven der Raumentwicklung" ("Perspectives of Spatial Development in Germany", BMVBS 2006b), though in the German text the term Leitbild is still used extensively.

Towards a new Type of Content

New efforts to redefine the meaning and substance of Leitbild had been under way for some years. The old definition, which had been contained in the federal law between 1989 and 1998, was apparently too vague to serve modern spatial planning. In an overview article of 1995 in the Handwörterbuch der Raumordnung Lendi had brought together the thinking of the time, including the critical points of vagueness and missing scientific basis. He defines Leitbild as a situation which is meant to be in the future and is to be aimed at by concerted action (Lendi 1995: 624). Ten years later in the next edition of the Handwörterbuch der Raumordnung Dehne follows a similar approach, mentioning the critical points of vagueness and non-scientific elements (Dehne 2005: 611-612). As to the argument of not being scientifically based, he adds quotations that this might even lead to mental manipulation and cites Theodor Adorno, who supposed that the concept of Leitbild "even has a soft military sound" (Dehne 2005: 612 on Adorno 1967).

Regarding content, Dehne explicitly mentions two items. One is the long-existing demand for similar living-conditions in all regions of the country. The second is sustainability, with all of its three "pillars", the economic, social, and ecological sustainability. Looking at this more closely, the similarity of living conditions is contained in the social aspect of sustainability, which in turn is in considerable conflict with the other two, especially the economic sustainability (Zimmermann 2003). Sustainability also had become the content of the Federal Law on Spatial Planning after its 1998 revision.

Much can be said against the vagueness of this new Leitbild (now called – synonymously – Leitvorstellung in the law), as can be said against the concept of sustainability as such (Zimmermann 2003). But at least it is connected to a wide range of literature based on environmental momentum, and it is no longer a semi-philosophical reasoning focused on spatial planning alone. – This three-fold sustainability is, last not least, at the same time the structural basis for the now valid version of the Leitbild documents.

Contents and Structure of the 2006 Leitbilder

Other contributions in this volume deal with individual elements of the new Leitbilder. Therefore, only some overriding remarks follow here. The first relates once more to definition. The basic document refers to the "Leitvorstellung" in the federal law and then, without further comment on definitions, moves on to three "Leitbilder", thereby connecting itself explicitly to a German tradition of having a Leitbild at some place near the top of the goal-instrument-hierarchy.

But what is called Leitbild in this document could as well be termed three objectives or goals, and this would simultaneously provide a clear meaning of the heading of that document. "Objectives and action strategies for the spatial development in Germany" would be easily understood without having to look up the word "Leitbild". The English reader is saved from this necessity, because in its unofficial translation of the heading of the basis document the Ministry calls it: "Concepts and Action Strategies for the Spatial Development of Germany" (BMVBS 2006b: 57). There are adequate German translations for "concepts", such as Konzepte or Konzeptionen, but the Ministry chose to connect itself explicitly to the German Leitbild tradition. Thus, the three Leitbilder of 2006 are in reality objectives, for which then certain instruments are sketched[10]. In the English translation (BMVBS 2006b: 38-55) the three "concepts" are named "Growth and innovation", "Ensuring services of public interest" and "Conservation of resources, shaping of cultural landscapes". They are to correspond to economic, social and ecological sustainability. As was shown above for the first concept, all three are based on thorough research on the status in that field, on driving forces etc.

What is important is that the concepts themselves are not disguised as scientific outcomes, and instead they openly constitute political intentions. The resolution to accept the basic text states that the group of ministers sees this as their political intention (BMVBS 2006a: 26-27). And the text for the three Leitbilder is expressed in terms like "spatial planning shall do this and that".

A major political decision is to be seen in the sequence of the three concepts. 20 or even 10 years ago it would have been unthinkable that the growth concept came first and the spatial equalization concept, which had been unchallenged in dominating German spatial planning for decades, followed as number two. This is the result of very recent re-thinking of spatial planning. Not least through the Lisbon process, Germany has started to think about the regional sources of national growth (Zimmermann 2007). The new approach also links concepts and instruments directly. Before, they had been dealt with separately in two documents, both accepted at different dates by the group of ministers (ORA (1992); HARA (1995)). The new document explicitly replaces both at the same time (BMVBS 2006a: 26).

2.5 Concluding Remark: „Leitbild" – what for?

The new version of Leitbild, as laid down in the basic document (BMVBS 2006a), constitutes a positive development from SARO and Dittrich as well as from later statements based on their ideas and often wording. Looking back it was a notion lacking clarity methodologically, originally formed by a scientist (E. Dittrich) alone, and then jointly by scientists and ministry delegates in the SARO group. It now has proceeded to an explicitly political statement of a political body, and at the same time it is based on solid empirical investigation.

The final question is, however, why the term Leitbild (or its equivalent of Leitvorstellung in the federal law) still seems necessary. Alright…maybe the vague notion of sustainability could be connected with this – equally vague - heading.

10 That change to the better had started earlier, namely with a document of the ministry of 1993 (ORA 1993). That had already gone from the single multi-faceted notion of a Leitbild to five separate Leitbilder for five policy fields like settlement structure, environment, transportation etc.

This does not do much harm, because more stringent discussions and any focused political program can do without it anyway, working instead immediately with concrete well-described subjects in a specific field. Whether or not these can then be subsumed under any of the three "pillars" of sustainability, is more a matter of word games than a direct derivation of any substance. The author has the impression that German spatial planning would lose close to nothing if the term Leitbild was replaced by "goals", "objectives", "perspectives" etc., to be followed by instruments, action plans etc. Scientists would then take these objectives as political statements and would, instead of pretending to provide them, simply discuss them – as this paper did with the term Leitbild!

References

Adorno T. (1967): Ohne Leitbild. In: Adorno T. (ed), Ohne Leitbild. Frankfurt am Main, 7-19.

Aring J. and Sinz M. (2006): Neue Leitbilder der Raumordnung in Deutschland. Modernisierung der Raumordnungspolitik im Diskurs. In: DISP, No. 165, 43-60.

BMVBS (2006a): Leitbilder und Handlungsstrategien für die Raumentwicklung in Deutschland. Submitted for approval of the Ministerkonferenz für Raumordnung, 30.06.2006. Berlin: Bundesministerium für Verkehr, Bau und Stadtentwicklung.

BMVBS (2006b): Perspectives of Spatial Development in Germany. Berlin: Bundesministerium für Verkehr, Bau und Stadtentwicklung. (German version: BMVBS (2006): Perspektiven der Raumentwicklung in Deutschland. Berlin: Bundesministerium für Verkehr, Bau und Stadtentwicklung.)

Brockhaus Enzyklopädie (1990) 19. ed., Vol. 13. Mannheim.Dehne P. (2005): Leitbilder in der räumlichen Entwicklung. In: Handwörterbuch der Raumordnung, 4. ed. Hannover: Akademie für Raumforschung und Landesplanung, 608-614.

Dittrich E. (1953): Versuch eines Systems der Raumordnung. Lecture, Mitgliederversammlung der Landesplanungsgemeinschaft Westfalen, 21.11.1953, Institut für Raumforschung, Vorträge, No. 4.

Dittrich E. (1962): Raumordnung und Leitbild. Wien.

Eucken W. (1952/1990): Grundsätze der Wirtschaftpolitik, 6. ed. Tübingen.

HARA (1992): Bundesministerium für Raumordnung, Bauwesen und Städtebau, Raumordnungspolitischer Handlungsrahmen. Resolution of the Ministerkonferenz für Raumordnung in Düsseldorf, 08.03.1995. Bonn.

Leendertz A. (2008): Raumforschung, Raumplanung und NS-Vergangenheit. – Forschungsgegenstand, Deutungen, Kontinuitäten. Manuscript.

Lendi M (1995): Leitbild der räumlichen Entwicklung. In: Handwörterbuch der Raumordnung, 3. ed. Hannover: Akademie für Raumforschung und Landesplanung, 624-629.

ORA (1995): Bundesministerium für Raumordnung, Bauwesen und Städtebau, Raumordnungspolitischer Orientierungsrahmen. Leitbilder für die räumliche Entwicklung der Bundesrepublik Deutschland. Bonn.

Puga D. (2001): European regional policies in light of recent location theories. CEPR Discussion Paper No. 2767. Toronto: Center for Economic Policy Research (CEPR).

SARO (1961): Die Raumordnung in der Bundesrepublik Deutschland. Stuttgart: Sachverständigenausschuss für Raumordnung.

Sinz M. (2005): Raumordnung/Raumordnungspolitik. In: Handwörterbuch der Raumordnung, 4. ed. Hannover: Akademie für Raumforschung und Landesplanung, 863-872.

Sinz M. (2006): Die neuen Leitbilder der Raumentwicklung. Anmerkungen zu einem politischen Diskurs. In: Informationen zur Raumentwicklung, No. 11/12, 1-8.

Storbeck D. (1982): Das „Leitbild" der Raumordnung. In: Grundriß der Raumordnung. Hannover: Akademie für Raumforschung und Landesplanung, 211-216.

Turowski G. (ed., 2002): Spatial planning in Germany. Structures and concepts, Studies in Spatial Development, Vol. 1. Hannover: Akademie für Raumforschung und Landesplanung.

Webster's encyclopedic unabridged dictionary of the English language (1989). New York.

Wurzer R. (1962): Laudatio, in: Dittrich, Erich, Raumordnung und Leitbild. Wien, 1-4.

Zimmermann H. (1966): Zielvorstellungen in der Raumordnungspolitik des Bundes. In: Jahrbuch für Sozialwissenschaft, Vol. 17, No. 2, 225-245.

Zimmermann H. (1994): Die Rolle von „Prinzipien" in der Theorie der Wirtschaftspolitik. In: Ordo, Vol. 45, 137-149.

Zimmermann H. (2003): Nachhaltigkeit im Raum: „Hart stoßen sich die Sachen". In: Ritter E.-H., Zimmermann H., Nachhaltige Raumentwicklung – mehr als eine Worthülse? Forschungs- und Sitzungsberichte, Vol. 219. Hannover: Akademie für Raumforschung und Landesentwicklung, 17-33.

Zimmermann H. (2004): Agglomeration tendencies and national growth. A primer on recent developments. Discussion Paper Series "Fiscal federalism and economic growth", No. 4. Marburg: Philipps-University Marburg, Public Finance Group (to be obtained from the author by email: horst.zimmermann@debitel.net).

Zimmermann H. (2007): Die Rolle „mittlerer Regionen" aus ökonomischer Sicht: Regionen zwischen Agglomeration und ländlich-strukturschwachem Raum. In: Köhler S. (ed.), Wachstumsregionen fernab der Metropolen, Vol. 334. Hannover: Akademie für Raumforschung und Landesplanung, 8-21.

3 Engines of Growth beyond Metropolitan Areas – Success Factors and Recommendations for Action to Support Development

Gabi Troeger-Weiß, Hans-Jörg Domhardt, Christoph Scheck

Introduction

With discussion on the "Concepts and Strategies for Spatial Development in Germany" put forward by the Standing Conference of Ministers responsible for Spatial Planning in Germany (the MKRO), the burgeoning importance of growth areas and metropolises as engines of growth and dynamic performers has become unmistakably clear. However, alongside these regions, there are also numerous rural regions and other more peripheral areas in Germany, located outside the more narrowly drawn boundaries of metropolitan conurbations which, from the perspective of regional economy and economic geography, are currently displaying positive and sustainable development, thus contradicting the more generalised impression of such regions as "areas of shrinkage and population depletion" (cf. BBR 2005: 85ff.). These areas exhibit major potential for growth in tandem with rising population and employment numbers and a high level of gross value added. These regions should also be made the subject of attention and of funding in the future, in the context of securing parity of living conditions in order to support existing positive trends.

Comprehensive spatial planning at the federal level has recognised the potential of such regions to contribute to macroeconomic growth. In the document "Concepts and Strategies for Spatial Development in Germany" (cf. Secretariat of the standing conference of state ministers with responsibility for spatial planning in Germany (MKRO) at the Federal Ministry for Transport, Building and Urban Affairs (BMVBS) 2006), which places the vision of "Growth and innovation" on a par with the other two visions of "Ensuring services of public interest" and "Conservation of resources, shaping of cultural landscapes", it is clearly stated that, even outside metropolitan regions, there are other regions and locations which make a major contribution to economic growth and display a distinct and sustainable development profile complementing the metropolitan regions (cf. ibid.: 12). Federal-level spatial planning also recognises the need to target resources on supporting such growth areas since they may well rank internationally as centres of innovation and of specialist technological excellence (cf. ibid.: 15). Moreover, the graphics illustrating this vision do indeed depict possible "growth areas outside the more narrowly drawn boundaries of metropolitan conurbations" (cf. figure below).

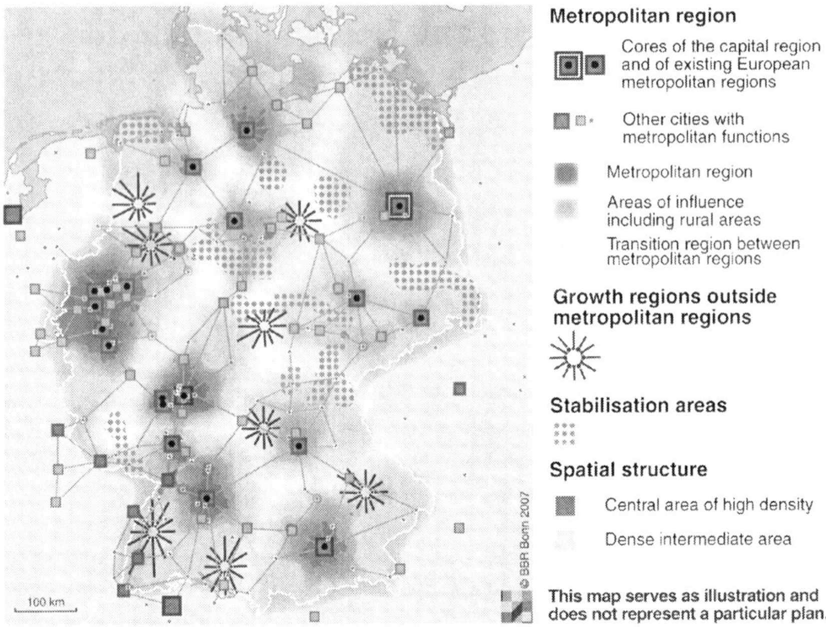

Fig.1: Leitbild "growth and innovation". Source: Secretariat of the standing conference of ministers with responsibility for spatial planning in Germany (MKRO) at the Federal Ministry for Transport, Building and Urban Affairs (BMVBS) (ed., 2006): Concepts and Strategies for Spatial Development in Germany, Berlin, 13.

What options exist for supporting and enhancing existing engines of growth, and for initiating similar developments in other regions? It is precisely to investigate such matters and to develop concrete proposals for developing and assisting such areas – in a reworking of the concept of endogenous regional development, so to speak – that the Federal Ministry for Transport, Building and Urban Affairs (BMVBS), working in partnership with the Federal Office for Building and Regional Planning (BBR) within the programme entitled "Demonstration projects of spatial planning", has called upon the services of the research project on "Factors underlying successful engines of growth beyond metropolitan areas".[1]

1 The project was undertaken between October 2006 and October 2007 at the chair of regional development and spatial planning at the University of Kaiserslautern by Prof. Dr. Gabi Troeger-Weiß, Dr. Hans-Jörg Domhardt, Christina Kaltenegger and Andreas Hemesath. The project team was supported by Jens-Uwe Staats of the Federal Ministry of Transport, Building and Urban Affairs (BMVBS) and by Dr. Rupert Kawka of the Federal Office for Building and Regional Planning (BBR). Detailed findings have been published (in German) in: Bundesministerium für Verkehr, Bau und Wohnungswesen (BMVBS) / Bundesamt für Bauwesen und Raumordnung (BBR) (ed., 2008): Erfolgsbedingungen von Wachstumsmotoren außerhalb der Metropolen, Werkstatt: Praxis, No. 56, Berlin/Bonn.

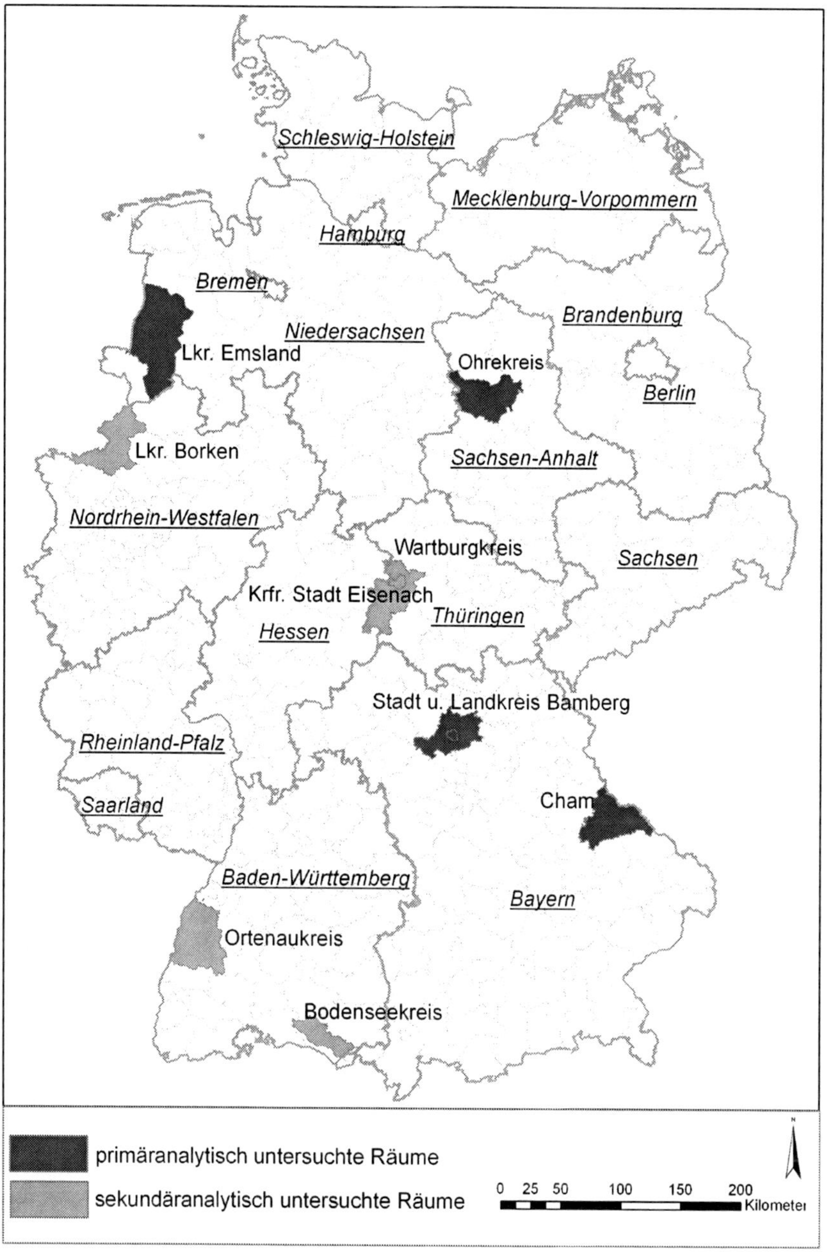

Fig. 2: Map of the study areas in Germany. The dark blue areas have been primarily analysed, the light blue areas have been secondarily analysed. Source: Chair of regional development and spatial planning of the University of Kaiserslautern 2007.

The aim of this project was to identify and describe the factors behind the emergence of dynamic corridors, smaller-scale agglomerations, medium-sized towns, and intensive-growth locations outside the more narrowly drawn boundaries of metropolitan conurbations. The regions selected in this context for closer analysis were the Bamberg region (Stadt und Landkreis Bamberg), Lake Constance County[1] (Bodenseekreis), Borken (Landkreis Borken), Cham (Landkreis Cham), Emsland (Landkreis Emsland), Ohre (more precisely Börde County, Bördekreis), Ortenau County (Ortenaukreis), and Wartburg County/Eisenach (Eisenach und Wartburgkreis), all of which are currently contributing in a disproportionately high manner to macroeconomic growth and are shown in the map below. This economic growth is defined primarily in quantitative terms. The second stage is to identify a scope for action on the part of comprehensive spatial planning and regional development capable of supporting and promoting the positive developments in these growth regions on a sustainable basis.

In terms of methodology, the first two parts of the study (theoretical fundamentals, structural analysis) are based on a survey of literature and documents, as well as analysis of the quantitative data supplied by the statistical offices of the federal states. The process of isolating the determinants and factors which account for the success of the engines of regional growth draws essentially from some 80 guided expert interviews conducted with representatives of the private sector, government, and other organisations located within the study area. This is supplemented by findings from a survey conducted among the general population in one of the two sub-regions. These three components subsequently provide a basis for deriving and identifying the factors which account for the success of engines of growth beyond metropolitan areas, for the conclusions which are drawn, and for the recommendations for action on the part of comprehensive spatial planning which are put forward. Owing to constraints on length, it has not been possible to deal with all aspects in full within the framework of this paper; readers are therefore advised to consult the published research report, which can be ordered or downloaded for free from the BBR (www.bbr.bund.de).

3.1 Basic Principles

To find a way of initially approaching the topic of "growth regions or engines of growth" from a theoretical point of view, a number of different explanatory approaches of variations in the dynamics of regional growth need to be considered. At this point it is useful to summarise the theoretical concepts associated with regional and regional-economic growth in regard to the task at hand, in order to evaluate existing approaches, programmes, and projects for promoting growth

2 NB: The „Lake Constance County" is not the county round the city of Constance but in fact the county round the city of Friedrichshafen at the north bank of the Lake Constance.

regions outside metropolises. Using the criteria thus derived, it will subsequently be possible to establish whether and in what way the various approaches and programmes providing funding in the individual growth regions represent, in theoretical terms, a suitable means of promoting growth in the region.

In terms of the concept of knowledge-based regional development (cf. Maier, Tödtling, Trippl 2006: 107ff.), it is in particular those programmes and projects aimed at establishing or strengthening regional network structures which are viewed positively. Whether this is within the framework of the "innovative milieu" (cf. Fromhold-Eisebith 1999: 30f.; Maier and Tödtling 2006: 89f.), a "regional innovation system" (cf. Thomi and Werner 2001: 202ff.), or a "learning region" (cf. Stahl and Schreiber 2003: 27, Hassink 2001: 222), these approaches are suitable means of promoting innovation – and growth along with it – within a region. These three theories of knowledge-based regional development lend themselves to transfer and application, by way of concepts, not only to scenarios of municipal agglomerations, but also to regions of a peripheral and more rural nature (cf. Fraunhofer-Institut für experimentelles Software-Engineering (IESE) 2006: 35f.). Consequently, some elements of these theories of growth are also applicable in the context of promoting engines of growth outside narrowly defined metropolitan conurbations. What really matters here as far as growth and innovation within a region are concerned, more so than geographical location, is human capital, especially in the form of a willingness and a capability to co-operate on the part of regional actors. Importance also has to be attached to other soft factors, such as everyday region-based awareness (cf. Danielzyk and Wiegandt 1987: 441ff.) and identification; measures aimed at raising awareness and strengthening the region's image (for example, by cultivating traditions) therefore offer good potential for the development of a region. One other important factor affecting regional development is the political structure: not infrequently, strong and committed personalities – especially if they enjoy widespread support and acceptance – are able to provide the critical ingredient which gives one region the competitive edge over other regions.

In those approaches which attempt to explain spatial development on the basis of polarisation theory (cf. Maier, Tödtling, Trippl 2006: 77ff.), it is also quite common to find references to the effects of image. In some cases, disadvantaged regions have to live with negative prejudices, misjudgements, and misinformation, all of which impede positive developments (cf. Hamm and Wienert 1990: 40). Accordingly, this is also a point at which growth strategies and assistance programmes should be applied; in other words, external communication beyond the region has an important role to play. Image campaigns within a regional-marketing strategy can make an important contribution to the positive development of regions located outside the metropolitan conurbations, and are therefore clearly to be viewed in a positive light. Projects and programmes aimed at improving public relations are particularly important instruments in growth regions with rural/peripheral locations characterized by a long history of lacking development.

Since growth regions outside metropolitan conurbations rarely contain dominant urban centres exerting agglomeration effects, it is possible only to a very limited extent to apply to these regions any theory based on growth poles and their spatial dimension, whereby a growth pole takes on a growth-promoting function in respect to its hinterland. Moreover, since it is easier for the sectorally diversified economic structures which – according to the sectoral cycle thesis – are so important for long-term growth to become established in larger regions (cf. Hamm and Wienert 1990: 37), it would appear highly probably that, in the case of those regions located at some distance from a metropolis and which frequently contain only so-called "intermediate-order centres", it is only the region in its entirety which could be capable of representing a growth pole.

Such other elements of infrastructure as airports or universities may also constitute necessary – though not sufficient – impulses for regional growth (cf. Frey 2005: 471). Important impulses also frequently emanate from particular large companies, especially where these are export-oriented (cf. Stiller 2005: 850). Ultimately, the critical feature is that the factors which support growth should be interconnected and put to use in the interests of the entire region. Comparisons can be made here with the network structures typical of knowledge-based regional development, which once again highlights the importance of this concept.

In this context, it is also important to consider regional strategies aimed at developing clusters (cf. Zürker 2007: 54f.); assistance measures to this end should also be viewed positively. Linking up business and research, bundling knowledge and competencies, and sharing infrastructure are all ways of bringing success to initiatives promoting the emergence of clusters; consequently, they all promote regional development. The region is rendered all the more competitive by follow-up effects, such as the appeal it exerts on new businesses deciding where to locate to, by the new jobs which are created, as well as by the various positive, long-term effects which emerge for the region. Consequently, economic clusters can also be regarded, to some extent, as powerful engines of regional growth, and thus as performing an important role outside metropolitan conurbations (in the more narrowly defined sense).

With regard to assessing existing approaches, programmes and projects aimed at promoting growth regions outside metropolises, and based on the theories considered, it is therefore possible to conclude that the following structural attributes may well be regarded as success factors, and therefore constitute the key criteria for the following study:

- regional network structures,
- intensive partnerships between and among regional actors,
- highly committed, strong personalities from the political and business arenas,
- initiatives aimed at strengthening regional image (e.g. regional marketing),
- organisations providing active assistance to support economic development (in particular to help small- and medium-sized enterprises),
- elements of infrastructure or enterprises capable of acting as "beacons",
- networking and co-operative structures, as well as partnerships for knowledge transfer, which foster clustering, and
- a high degree of orientation towards export within the regional economy.

3.2 Development to Date – Viewed from a Structure-Analysis Perspective

In order to filter out any initial parameters and factors which might either promote growth or possibly impede development, a structural analysis was carried out, first of all in the fields of spatial structure and settlement structure, population, economy and labour market, and infrastructure. In the following we present the key strengths and weaknesses, as well as growth-promoting and growth-impeding structures, which have emerged as the central findings of the structural analysis. This makes clear just which factors are decisive for all eight – or at least for most of the eight – researched areas, or, alternatively, are specific to just a few individual areas (cf. BMVBS/BBR 2008: 11ff.).

Spatial Structure and Settlement Structure

One particularly characteristic feature of all of the regions studied is the capacity for further provision with higher-level central-place functions; this must equally be regarded as a factor which impedes development. Six of the eight study areas do not contain a single higher-order centre; in Ortenau County and the Bamberg region, the only higher-order centres, Offenbach and Bamberg, have populations of around 60,000 and 70,000 respectively. Moreover, five of the eight study areas are also characterized by being located in the periphery: only the Wartburg and Ohre region and the Bamberg region are not located adjacent to a national border. Today, such a marginal location may well be viewed as advantageous; in the past however, when borders were less porous, such a location tended to be seen as a disadvantage. With regard to population density, it can be noted that in six of the eight study areas settlement can be described as sparse, an additional factor impeding development. Only in the counties of Lake Constance and Borken is population density above the national average.

Population Structure and Development

Turning to population development, what is particularly striking is the largely positive trend recorded over the last 25 years; this trend would then tend to rank as a factor fostering development and growth. Over this period, population loss has only been recorded in the two regions in eastern Germany; only in the Wartburg/Eisenach region on-going and longer-term shrinkage is in process. With the sole exception of Cham County, where population development levelled off more than in the state of Bavaria as a whole, all of the study regions can look back at a pattern of population development more positive than their respective federal state. Equally with regard to population projections for the future, with the sole exception being Cham and the Wartburg region, all of the study areas can expect population development at a rate above the national average, resulting in growth in population size. Positive population development will therefore remain a strength of growth regions located outside metropolitan conurbations in the future, as such areas will not be confronted to the same degree, or with the same urgency, with the problems associated with demographic change.

A further positive feature of these areas is their age structure. The regions analysed generally tend to be marked predominantly by their relatively young populations – an additional factor to be noted as fostering development. Only in the two study areas in eastern Germany do below-20-year-olds comprise a smaller proportion of the total population than the national average. Accordingly, with regard to population structure and development, it is also only these two areas – and to some extent also Cham county – that exhibit specific negative features. Whilst all the other regions can both look back at and look forward to continuous population growth, and equally to a relatively "young" population structure, the Wartburg region, with unmistakable decline in numbers and a relatively old population structure, is particularly badly positioned.

Economic Structure and Development

Without exception, and with regard to economic structure and development, all of the study areas are characterised by the important position occupied by manufacturing. In all of the study areas, this sector accounts for a significantly higher proportion of the workforce than the national average; at the same time, however, this underlines the below-average significance of the services sector in these regions. In contrast to developments at state level, over the last 25 years the number of people employed in manufacturing in many of the study areas has fallen either only slightly or not at all. In summary, therefore, it should be noted that all of the growth regions analysed share one common distinctive feature in the strong industrial base that each one boasts. In terms of employment figures, the development has also been generally positive; since the mid-90s, only the two counties in eastern Germany have recorded a decline. By contrast, since 1980 the other six counties have achieved high rates of growth in employment overall; in all cases this exceeded the state average. In this regard, even the two regions in eastern Germany recorded a better development than the federal state as a whole. This is the reason why positive employment development can be cited as a distinguishing feature of the growth regions studied, and equally can be viewed as a factor which will foster development in the future.

With the exception of the two areas located within the new federal states (the former East Germany), where it has only been possible to analyse sectoral developments since the end of the 1990s, the high rate of growth in the services sector is a further key feature in all of the areas surveyed. Admittedly, the baseline for growth in this sector was relatively modest. What this development reveals is that in all of those regions which are still characterised by the relative strength of the secondary sector, sectoral restructuring has left its mark in the meantime; the tertiary sector is now also a major contributor to the positive development recorded in these regions.

In most of the areas studied, the sectoral structure of the economy can be seen as one which fosters development. Most of these areas display a particularly high degree of diversification in regard to sectoral structure, with primarily small- and medium-sized enterprises. The exceptions in respect of diversification are, however, the Wartburg region, which is dominated by its metal-working and vehicle-construction sector, the Lake Constance area, with a strong machine-construction sector, and Cham, where a particularly large number of people are employed in the electronics sector. The Wartburg/Eisenach region is marked in particular by the very heavy concentration on a small number of sectors; this is a specific factor which has the potential to impede the long-term development of the region. Consequently, in terms of economic development and economic structure, it is also necessary to note that the Wartburg region exhibits only very few of the features which typify the other growth regions. These are, generally speaking, marked by very positive development in regard to employment, a stable secondary sector, a flourishing tertiary sector, and a broadly diversified economic structure characterised by small- and medium-sized enterprises. In regard to unemployment, the Wartburg region and the Ohre region stand out as the only two of the regions studied to record rates above the national average for Germany. What sets the other regions apart, by contrast, are unemployment rates which lie well below the average for the state in which they are located; this is a key feature of engines of growth outside metropolitan regions.

Infrastructure

With regard to infrastructure, six out of the eight researched areas are particularly well endowed in terms of transport infrastructure; more specifically, they have extremely good road networks, which incontrovertibly ranks as a positive and key factor contributing to development in these areas. It is solely the counties of Cham and Lake Constance that reveal a critical weakness in this regard. However, with the sole exception of Ortenau, all of the areas studied reveal weaknesses in regard to air-traffic infrastructure. Another inescapable and common deficit of all of the areas studied is the low level of provision of higher-grade educational establishments and research institutes, the most striking weak point of these regions. In this context, the university located in the Bamberg region allows it to claim a specific strength.

3.3 Determinants and Factors behind the Success of Regional Growth Engines

The following chapters describe the determinants and factors which account for the success of engines of regional growth. We begin with a description of the historical situation which forms the starting-point for growth, before moving on to consider technical determinants. This, in turn, is followed by a discussion of the determinants which affect regional adaptability, of organisational and institutional determinants, and of the person-related determinants (such as creative milieus).

The historical Starting Point and path Dependency

The key factors underlying the positive development which has taken place in the areas investigated are, primarily, the opening up of borders within Europe with the creation of the single European market within the EU-15 states in 1993, and equally the collapse of the Soviet Bloc in 1989, and the subsequent reunification of Germany in 1990. One further factor affecting European integration is the EU's enlargement to the east which took place in 2004. By way of illustration, until the collapse of the Soviet Bloc, the Bamberg region was relatively close to an international border; reunification has now given it a central location along the Munich-Berlin axis. After years of suffering quite severely from the disadvantages of their peripheral position on the fringes of the Federal Republic of Germany, Emsland, Borken and Lake Constance/Ortenau all benefited from the central locations they acquired within the EU as a result of the creation of the single European market.

Moreover, particularly in the years immediately following the end of the Warsaw Pact, Emsland recorded strong population growth, due largely to the influx of "resettlers" from formerly communist Eastern Bloc states. Between 1988 and 1996, over 20,000 such resettlers moved to Emsland, primarily from Russia and Kazakhstan. More generally, the development of the Emsland was also boosted significantly in the 1950s by the influx of refugees who settled in the region, and by the territorial demands made by the Netherlands in the wake of the Second World War. One effect of these European-level influences was the initiation of the so-called "Emsland Plan", which in historical terms was to constitute the single most important instrument affecting the positive development which has taken place in the county of Emsland. It is thanks to this comprehensive programme for development that the region has been able to cope with the various phases of economic restructuring without major problems.

In the case of the two study areas located in eastern Germany, it is German reunification which has proved to be the most significantly influential factor over the last two decades. The Ohre region in particular has profited greatly from this change at the national level. However, with the shift in its locational status from marginal to central, the Wartburg region has also been a major beneficiary and now ranks among the most prosperous regions of the state of Thuringia. The key factor for both areas was their location along a main transport route between eastern and western Germany.

In 2004 the EU's enlargement to the east finally allowed the county of Cham to profit from its location in close proximity to the markets of eastern Europe. Prior to the collapse of the Soviet Bloc, Cham occupied an extremely peripheral location at the very edge of western influence; it now finds itself much closer to the centre of the European Union. This has led to improved economic relations with the Czech Republic. For example, agreements on co-operation have been entered into between economic-development agencies, and a joint business portal has been set up.

A further influential factor at national level is the decline of agriculture. Especially in those areas which previously were shaped by agriculture, such as the Ohre region, Cham, and Emsland, solutions had to be found to compensate for the large number of jobs lost in this sector. Whilst Emsland's industrial sector has been able to create replacement jobs, the Ohre region has had much greater difficulty in restructuring its economy.

In addition to factors of influence on European or national levels, there are also long-term effects in individual areas which emanate from changes in the structure of the regional economy. For example, in Borken, and to some extent also in Emsland, the decline of the textile industry had severe negative impacts initially, especially in terms of labour-market structure. The consequences were similar following the job losses in the oil industry in Emsland. Very quickly, these regions witnessed a process of industrialisation which was both more advanced and of higher value, and which was thus able to compensate for the sharp decline in employment in agriculture. In the Wartburg region by contrast, the metalworking sector has been the key factor affecting development for decades. The role played for many years by mining as a source of work in the metal-working sector has more recently been taken on by the automotive industry. The consequences of a decline in this sector could be fatal for the region.

The city and county of Bamberg, another area subject to research, finds itself in a similar position. For many years, the local economy here was shaped by the ceramics industry; following the collapse of this industry, a monostructure was established around the automotive industry and car-component suppliers. The experts consulted in the course of the research project see a danger threatening the future of Bamberg here. One factor, however, which is consistently cited as critically affecting the positive development Bamberg has experienced, is the city's traditionally important location on the historic transport and economic axis formed by the valley of the Main and Regnitz rivers. For many years, Cham suffered from the peripheral position it occupied, not only in respect to Europe and Germany, but also Bavaria. However, this region, traditionally shaped by its agricultural and building sectors, also enjoyed one advantage attributable to the low level of settlement density, namely the speedy availability of as many commercial sites as were – and continue to this day are – needed.

Technical Determinants

In the present context, the term "technical determinants" of regional development includes diversification of the economic structure, the tourism industry, size-structure of businesses, the existence of cluster- and knowledge-based structures, as well as the general level of training and qualifications found in the workforce and the level of infrastructure provision. To sum up, although many of these technical determinants exist in a manifestation which fosters development, and thus may well impact favourably on regional development, only very few of these factors are to be regarded as basic prerequisites for the emergence of growth engines outside metropolitan conurbations. In respect of all of the technical determinants mentioned, there certainly are regions which break the pattern and, despite the presence of structures which manifest themselves negatively, nonetheless exhibit clear signs of growth.

The only technical determinant indicating positive development, and which thus can be seen as an important and key factor for success, is the size of businesses, in this case small- and medium-sized enterprises. However, what matters here is less the actual number of businesses in this category than the fact that, in regard to mentality, working practices and product range, SMEs in the researched areas tend to have developed profiles which are particularly conducive to supporting success.

Although the other technical determinants prove to be supportive of growth in several of the areas studied, other cases simultaneously demonstrate that these points are not in themselves critical factors for regional growth. To take one example, a highly diversified economic structure and the presence of corporate headquarters within the region may well be seen as a major advantage for long-term development, and yet other regions have shown positive development, despite their distinct concentration on just a small number of different sectors and the presence mainly of branches and subsidiaries rather than headquarters. The Bamberg region is a case in point. The same applies in respect to the level of provision of transport infrastructure. Although, in many regions, good transport links are a critical factor for regional development and clearly prove to rank among the factors which bring success, other study areas, such as the counties of Cham and Lake Constance, show that they should by no means be seen as necessary conditions for regional growth. In regard to the knowledge-based structures widely held to be critical for regional development, and which include a range of high quality educational and research establishments, only one of the areas studied (Bamberg) can claim to meet this standard to any significant degree.

It should be clear from the observations above that the mere existence of certain structures is not in itself sufficient to guarantee that a region will achieve a positive development; it is, rather, only through intensive co-operation and active, innovative management of the interplay among a host of different determinants that existing weaknesses can be cushioned or mitigated. For example, one feature which most of the areas studied have in common is an extremely active labour-market policy at local-authority level (e.g. training days in Cham County) coupled with a high level of willingness on the part of business to take on trainees: this is a critical contribution towards offsetting the detrimental effects which would otherwise ensue from a relatively low level of provision in the fields of education and research. In addition to this, in many regions a great deal of importance is attached to informal – but nonetheless intensive – corporate networks and to exploiting synergies; this is also a reaction to maximise the possible advantages which arise from the strategy of fostering economic clusters.

Ultimately, this demonstrates that it is not solely the "hard" locational factors (such as transport infrastructure, etc.) which are the determinants of success in growth regions, but, rather, and to a very significant degree, what are considered "soft" factors, such as the willingness to co-operate and the capacity to act on the part of regional players. The critical factor behind positive development is the success with which existing potential is exploited to compensate for any possible weaknesses.

Determinants associated with Regional Adaptability

The ability of a region for adapting is a key determinant of development. Thus, with regard to enabling regions to provide for future development, what is particularly important is the ability of a region to rise to the challenges associated with the sectoral restructuring of the regional economy, the ability to target the available subsidies, and the readiness to develop projects designed for promoting growth. Especially against the background of unmistakable shifts within the contextual parameters which have taken place throughout Germany due to demographic change, globalisation, and increasing competition among regions to attract inward investment, there is now a need for regions to come up with ever more new and innovative strategies if they wish to remain successful.

Except in the case of the areas studied in eastern Germany, the use of subsidies or funding is not a major issue. There is, for example, no perceivable link between the level of subsidy received and a region's success. Nevertheless, all of the areas studied claim subsidies and funding from the EU and from federal and the respective state governments to assist them in implementing projects and other measures. It is only partly possible to explain differences in the levels of federal funding claimed. The study areas located in Bavaria, for example, tend to rely more on state funding than on support from the federal government, as the procedure for applying for state funding appears to be somewhat less complicated. However, all of the experts consulted stress that external funding is never decisive in determining whether to proceed with a project or similar measure. This view could of course be influenced by the positive personal or insider perspectives of the actors involved.

Some conclusions about the priorities set for regional development, and equally on the level of innovation to be found in a region, can be drawn from the nature and diversity of regional projects and initiatives (e.g. the Franken River Paradise; the cross-border Ems-Dollart region; the Wartburg region's Learning Landscape, etc.). In the majority of regions, the county manager's office plays a crucial role in both the initiation and implementation of projects. Such projects promote the image of the county (or unincorporated city) both externally and internally. The importance of self-image can be measured by gauging familiarity with the project both among the general public and on the part of the experts consulted. The counties that score particularly well here are Cham and Emsland, both of which can boast a broad and diverse range of projects; the county of Ohre, by contrast, still has a lot of catching up to do.

All of the experts consulted agree in regarding networking between the business sector and scientific establishments, and also research and development, as a key factor for both growth and positive development in a region. Sustainable economic structures are more likely to emerge from this kind of symbiosis. Consequently, in most of the researched areas considerable importance has been attached to establishing business start-up centres and centres for innovation, and targeting the aid available to support economic development into building such networks. Among the main beneficiaries of such close co-operation with research institutions, universities, and other training establishments are the emergence of economic clusters, regarded by many experts as among the most sustainable ways for the economy to develop.

Organisational and Institutional Determinants

The organisational and institutional determinants reflect the capacity for innovation, the adaptability, and the commitment displayed by the actors within the region. To take one example, this study has been able to detect a correlation between the type and nature of initiatives to support regional development on the one hand, and the incidence of the other success factors which lead to regional growth.

Regional development initiatives hold an important position as innovative and – in most cases – informal organisational structures. They are better equipped to react to new demands of the kind that classical and formal organisational structures are frequently not suited for dealing with, or at least for reacting to quickly. Moreover, they directly target, and, respectively, enhance or redress the potentials and the weaknesses found within the areas studied. They therefore have a very important role to play within the researched areas, which is why many experts describe them as success factors for regional growth. Such initiatives are particularly abundant in Cham and Bamberg, with the county of Ohre still lagging well behind in this respect. Not only does Ohre have fewer development initiatives in place, but there is also little awareness of these among the general public.

A similar level of importance is attached to inner-regional networks. These have the function of bundling and focusing competencies, bringing regional actors into contact with each other, and promoting dialogue within the area concerned. Particularly important here are not only corporate networks, but also networks encompassing the private sector, politicians and public administration. These are viewed as the very foundation for creating a business-friendly climate. Experts have repeatedly stressed the importance of functioning networks both for attracting new businesses to locate in the region and for encouraging start-up businesses. One important factor affecting the success of functioning networks is certainly the size of the area covered. If this area is too large, then it becomes difficult to assemble the whole range of regional actors around one table. This probably explains why Cham stands out as such a positive example. The county's relatively small size makes it easier for the actors concerned to get to know each other and thus develop the willingness to create networks.

In addition to such networks, which bring together as many actors as possible, there is no dispute about the decisive impact on positive development in the areas studied attributable to leaders, trendsetters, and opinion formers. Their main contribution is as a central force which inspires others to follow. Such figures also serve as hubs providing a nexus and enabling them to bundle and co-ordinate diverse activities throughout the region. In most cases, such figures assume the position of county manager, and in this context special mention should be made of the present incumbents in Emsland and Cham. Again and again, however, people mention entrepreneurs who are particularly committed to their regions and actively strive to support them. Experts regard such "influential individuals" as among the key factors determining how successfully a region develops.

Personal Determinants – Existence of Creative Milieus

In respect of all of the areas studied, there is general agreement on the importance which is attached to identification and solidarity on the part of the general public and regional actors as a major factor for growth. According to the experts consulted, all of the areas studied enjoy high levels of identification with the region. The self-confidence attributed to the populations of the regions studied has its roots in the positive development which these areas can look back on, and which sets each one apart from neighbouring regions, the development of which may have been less positive. Especially in Cham and Emsland, reference has frequently been made to the high levels of regional solidarity displayed there by local businesses – most of which originate in the regions where they are now located. This factor in turn impacts positively on the structure of the economy in its entirety and helps to stabilise it.

In order to verify the comments made by the experts, wide-ranging surveys were conducted among the general public based on written questionnaires in four of the researched areas. In addition to inviting respondents to comment on such issues of quality of life and solidarity towards the region, the survey also addressed familiarity with projects, activities, and development initiatives within the respective study area. The results reveal that the various personal determinants always turn out to be particularly strong or weak in the same areas. For example, the Bamberg region records the highest levels of regional identification and solidarity, followed closely by Cham and Emsland. A similar picture emerges in respect to the self-confidence displayed by the public and by regional actors, self-image structures, and assessments of the factors affecting growth. On all of the factors covered, the Ohre region scored worst, although it should be added that, in all four of the areas subjected to more intensive investigation, the population to some extent "sensed" a more positive development in respect of a number of factors than could be justified by the collected structural data alone. This popular survey demonstrated that popular opinion in most cases does indeed coincide with the opinions of the experts consulted. A positive self-image of, identification with, and solidarity towards the region on the part of the general public, coupled with self-confidence on the part of both the public and regional actors, can make a positive contribution to regional development.

3.4 Factors Responsible for the Success of Growth Engines beyond Metropolitan Areas

In regard to the organisational and institutional determinants accounting for the success of engines of growth outside metropolitan conurbations, prime importance has to be attached to inner-regional networks, and equally to "influential figures" from politics and business. In all four of the areas subjected to closer investigation, the ways in which these two closely related phenomena manifest themselves are the key factors underlying positive development. The creation of informal networks bringing together business, politics, and public administration has generated a business-friendly climate in the areas studied, which is an extremely important factor for fostering development by attracting inward investment. Within these networks, it is in all cases the region's leading political players who act as the driving forces and key figures promoting integration. In Emsland and Cham, and also in the Ohre region and in the Bamberg district, it is the county managers or mayors who play an instrumental role in driving forward the development of the county or region they head. Through high levels of commitment, intensive and active support for economic development, and through their innovative ideas, a host of other regional actors also contributes significantly to the success of these regions. In most cases, clear majorities lend political leaders broad public support; this has in turn allowed these regions to develop a quite distinctive mentality. This is particularly tangible in Emsland and in Cham. Proceeding from an initial situation that was extremely difficult, both regions have witnessed the emergence of a community spirit and a mentality characterised by hard work, high levels of civic involvement, modesty, and strong identification with the region. Here, the hardships of the past have left people accustomed to working particularly hard. Considering the factors which account for the success of engines of growth outside metropolitan regions, it is these soft locational factors which come first.

By no means necessary conditions, but certainly helpful in bringing about positive regional development are specific manifestations of the technical determinants concerning economic structure and infrastructure. The way in which the economy is structured in the majority of the areas studied – characterised by a high level of diversification and a strong base of small- and medium-sized businesses, in many cases highly industrialised – makes these regions less susceptible to economic or structural crises, and renders them less dependent of the decisions of a small number of major corporations. A further characteristic feature of regional economic structures is the widespread presence of highly specialised market leaders for rather narrow niche markets and products, many of which are highly innovative and influential. A common feature of all of the areas studied is the extremely business-friendly climate they have created. Almost all of these areas also boast extremely good transport infrastructure. Good access to motorways is frequently seen as a particular key factor for success.

However, as is illustrated by the counties of Cham and Lake Constance, it is also possible to achieve positive regional development even in the absence of this particular advantage. What is particularly important once again is rather the wealth of endogenously initiated projects and the intensive work within networks already referred to. Particularly looking into the future, regional development initiatives as innovative organisational structures will rank as important factors for the continued positive development of regions situated outside metropolitan conurbations. There are positive benefits for a region from positioning and marketing itself as a self-contained growth region, from concentrating on its own particular strengths, and not depending on assistance from outside.

In summary, although there are some aspects of the factors referred to which are partly responsible for success in all of the regions studied, it can be stated at this point that in many cases there are also historically rooted, "one-off" features which make an important contribution to development. First and foremost is the nature and quality of the type of mentality referred to above. Development is also influenced by traditional or religious attitudes, for example by the Catholic mentality which predominates in the counties of Borken, Bamberg and Emsland, and which also contributes to the high rate of population growth which has taken place there. Traditionally strong industries, such as the metalworking sector in the Wartburg region, also shape long-term regional development to some extent. A further very specific situation arose in the Ohre region, as the area had the opportunity to profit from the problems suffered by the state capital Magdeburg following German reunification, when many investors relocated to the Ohre region.

Nevertheless, as a generalisation it is justifiable to claim that in all cases successful development is only possible when the potential that exists within a region is actually exploited by regional actors, and key decision-makers manage to exert a positive influence on development in the region through acting innovatively and adeptly. Exploiting the existing potential of a locality is also of fundamental importance when the inherent locational merits offered by rural areas are in play. In contrast to urban agglomerations, such areas – in most cases, at least – have the major advantage of possessing available land for development on a large scale, at relatively low cost, and with an attractive environment for living. However, a further critical point concerns regionally specific decision-making and development-control structures or specific milieus, which exist in only very few non-metropolitan regions and which would be necessary to help these regions to develop into growth engines.

3.5 Conclusions and Recommendations for Action at the Level of Comprehensive Spatial Planning

As far as the feasibility of implementing measures and strategies, and projects as well, is concerned, the following preconditions need to be in place:

- person-related structures,
- organisational structures.

In this context, special importance needs to be attached to the willingness of regional actors to bring to bear above-average levels of commitment for achieving joint regional development goals. Turning a peripheral, structurally weak, rural region into an engine of growth calls for

- structured conceptual groundwork,
- a high degree of regional self-confidence and determination,
- staying power,
- the ability to win the support of various groups within society, as well as that of actors at regional and local-authority levels, in the pursuit of joint objectives and strategies.

In addition to person-related and organisational requirements, a conceptual basis and a strategy for regional development are necessary. As many actors as possible should be able to identify with such a development strategy. In order to ensure the involvement and integration of a broad range of different groups within society, as well as policy-makers, the following approaches should be pursued:

- actions targeted on raising identification with the region,
- encouraging and strengthening honorary involvement,
- creating attractive living and working conditions,
- creating and supporting family- and senior-citizen-friendly structures (e.g. care services, supply infrastructure, attractive living environments, etc.),
- creating attractive educational infrastructure,
- providing places for vocational training in sufficient number and of an appropriate quality,
- support for finding suitable training positions for graduates,
- setting up vocational training initiatives to address, inform, and support students before they graduate, and to dovetail the needs of businesses with the interests of trainees,
- initiatives to integrate older workers within the workforce.

Other important prerequisites for developing and supporting engines of growth are

- the designation of high-grade industrial and commercial sites,
- providing these sites with the appropriate infrastructure,
- speedy and flexible administrative and approval procedures,
- support for new enterprises and spin-offs,
- investment and location marketing, as well as regional marketing, to attract inward investment and new industry,
- a business-friendly climate at local-authority and regional levels, in both political and administrative circles,
- the construction of networks for bringing together private enterprise, local-authority and regional policy-makers, professional and trade associations, etc. (e.g. competence networks, clusters, etc.).

The use of regional, locational and investment marketing as a means of creatively presenting the entire array of the region's strengths and potentials both to the local population and beyond the region receives special importance. The effects of this new instrument of superordinate federal and state-level planning are felt in the way they influence self-image and external image, particularly important in structurally weak rural areas. In the researched regions, regional, locational, and investment marketing have clearly and demonstrably contributed to:

- motivating businesses to stay at their current locations (maintaining the economic base),
- attracting new businesses (start-ups, relocations) and
- winning over sections of the population to support the region.

In summary, there is a need for a package of different actions to secure positive development in both terms of regional economics and the region itself.

As far as the question of transferring these findings to other regions is concerned, this is certainly possible, albeit only where the relevant conditions outlined above are in place. Essentially, these approaches and strategies for development can be seen as an orientational basis; the study illustrates ideas and provides examples of best practice. What has to be taken into account is that processes of change – such as those associated with socio-economic restructuring – are on-going processes. As the researched regions demonstrate, the journey from being a structurally weak region to becoming an engine of growth usually lasts at least ten to fifteen years.

References

Bundesamt für Bauwesen und Raumordnung (ed., 2005): Raumordnungsbericht 2005. Report, No. 21. Bonn.

Bundesministerium für Verkehr, Bau und Wohnungswesen / Bundesamt für Bauwesen und Raumordnung (ed., 2008): Erfolgsbedingungen von Wachstumsmotoren außerhalb der Metropolen, Werkstatt: Praxis, No. 56. Berlin/Bonn.

Danielzyk R., Wiegandt C.-C. (1987): Regionales Alltagsbewusstsein als Faktor der Regionalentwicklung? – Untersuchungen im Emsland. In: Informationen zur Raumentwicklung, No. 7/8, 441-449.

Fraunhofer-Institut für experimentelles Software-Engineering (ed., 2006): Wissensbasierte Regionalentwicklung – Diskussion der Bedeutung außeruniversitärer Forschungseinrichtungen für den Transfer von Wissen und Technologie in kleinere und mittlere Unternehmen (KMU) – Das Beispiel des Fraunhofer-Instituts für experimentelles Software-Engineering in Kaiserslautern. Kaiserslautern.

Frey R. (2005): Infrastruktur. In: Akademie für Raumforschung und Landesplanung (ARL) (ed.): Handwörterbuch der Raumordnung, 469-475.

Fromhold-Eisebith, M. (1999): Das „kreative Milieu“ – nur ein theoretisches Konzept oder Instrument der Regionalentwicklung? In: Raumforschung und Raumordnung, No. 2/3, 168-175.

Hamm R., Wienert H. (1990): Strukturelle Anpassung altindustrieller Regionen im internationalen Vergleich, Schriftenreihe des rheinisch-westfälischen Instituts für Wirtschaftsforschung Essen, No. 48. Berlin.

Hassink R. (2001): The Learning Region – A Fuzzy Concept or a Sound Theoretical Basis for Modern Regional Innovation Policies? In: Zeitschrift für Wirtschaftsgeographie, Vol. 45, No. 3/4, 219-230.

Maier G., Tödtling F. (2006): Regional- und Stadtökonomik 1 – Standorttheorie und Raumstruktur, 4th updated and amended edition. Wien/New York.

Maier G., Tödtling F., Trippl M. (2006): Regional- und Stadtökonomik 2 - Regionalentwicklung und Regionalpolitik, 3rd updated and amended edition. Wien/New York.

Secretariat of the standing conference of ministers with responsibility for spatial planning in Germany (MKRO) at the Federal Ministry for Transport, Building and Urban Affairs (BMVBS) (ed., 2006): Concepts and Strategies for Spatial Development in Germany. Berlin.

Stahl T., Schreiber R. (n.f.i.): Regionale Netzwerke als Innovationsquelle – das Konzept der „lernenden Region“ in Europa, Campus Forschung, No. 868. Frankfurt/New York.

Stiller S. (2005): Raumentwicklung, ökonomische. In: Akademie für Raumforschung und Landesplanung (ARL) (ed.): Handwörterbuch der Raumordnung, 850-856.

Thomi W., Werner R. (2001): Regionale Innovationssysteme – zur territorialen Dimension von Wissen und Innovation. In: Zeitschrift für Wirtschaftsgeographie, Vol. 45, No. 3/4, 202-218.

Zürker M. (2007): Cluster als neue Komponente der wirtschaftsbezogenen Raumentwicklung, Dissertation, Materialien zur Regionalentwicklung und Raumordnung, No. 22. Kaiserslautern.

4 Integrated Leitbilder for Urban Development Policy: The Example of the University and Hanseatic City of Greifswald

Thomas Kuder

Located in the sparsely populated northern German state of Mecklenburg-Vorpommern, Greifswald is a medium-sized city located by the Baltic Sea, close to the Island of Ruegen. The city's history as a member of the Hanseatic League and long established university is today reflected in its official title: The University and Hanseatic City of Greifswald.

Since the political and social upheaval following the dissolution of the German Democratic Republic in 1989, Greifswald, as many other German cities, has been in a continuous state of socio-economic, demographic change and has experienced significant transformation in its urban fabric and structure. These changes have been most clearly visible in the drastic decrease in both jobs and population, and can be traced back not only to post-socialist transformation processes, but also to processes of restructuring that are associated with an increasingly globalized economy.

While the cities of the western industrial world have been experiencing economically driven changes of both their social and physical structures for many years, eastern German cities have only begun to experience these changes following the revolutionary upheaval caused by the fall of the GDR. Therefore, the changes experienced by eastern German cities have been both more intense and have occurred in a much shorter time. In addition, these cities have also had to deal with problems that originated during the time of the GDR government, such as architectural decay and extreme environmental degradation.

Despite the signs of urban decay and increasingly acute decline in the number of jobs and population in Greifswald, since the early 1990s the idea of growth has defined local urban development policy. Thus, the various actors and institutions that influence urban development in Greifswald — including official plans and urban development programs —, have focused on promoting economic growth while ignoring the underlying causes contributing to economic and demographic decline.

Urban development policy, as it was practiced and reflected in various plans, designs, and suggested urban development projects, continued to anticipate significant growth in both the population and the economy. Despite the fact that this expectation of growth was contradicted by the realities and problems of the time, urban growth nevertheless remained the defining paradigm for urban development in Greifswald for over half a decade.

It was only a clear lack of local economic growth reflected in an absence of job growth that led to a change of course in the second half of the 1990s. The clear failure of growth-oriented urban development strategies caused local actors to begin to seriously question related strategies and policies — based on "tried and true" paths of commercial and industrial development — that had guided urban development until that point. In addition, an increasing awareness of the changing external conditions contributing to growth also led to scepticism towards growth-oriented strategies.

In response, and as part of creating a new city marketing process, a new direction for urban development was sought. At the centre of this new approach was the development of a paradigmatic approach, including the development of an overall concept, or "Leitbild" as it is called in German, for urban development, which occurred between 1996 and 1999.

Using Greifswald as a case study, the following paper demonstrates how the process of creating a new model for urban development can influence its practical implementation. A reconstruction of this process will illuminate the opportunities and limits of creating an overall urban development concept, as well as the impact of the integration of this overall concept into official policy on the specific urban development challenges faced by Greifswald.

First, the various stages of urban development in Greifswald will be described, including an illustration of the evolution of the corresponding institutional framework for urban development policy and institutional development pathways. Thus, the effects of Greifswald's specific institutional context will be explained. Second, a description and analysis of the process of creating a new overall concept, as it reflects an institutional "turning point", i.e. a process of instigating institutional change, will also be assessed in terms of its effects on urban development policy.

A few excerpts of interview material are also presented here. Although they have been reported in an anonymous fashion (marked as: Greifswald Interview HGI, interview number: line number of transcription), these citations, stemming from key-informant interviews with important actors influential in the urban development of Greifswald, illustrate the attitudes that characterized this process. These interviews were held by the author between 2004 and 2007 as part of the research conducted by the IRS Department 4, for "The Regeneration of Shrinking Cities."

Finally, the meaning of the overall concept for urban development policy and practice will be analyzed. This analysis will highlight the various structural and institutional development processes that enabled the creation of a new approach towards urban development. On the basis of a theoretical discussion, this analysis will allow for general conclusions regarding the importance of the creation of an overall concept for urban development policy.

4.1 The University and Hanseatic City of Greifswald

The original settlement of Greifswald can be traced back to the founding of a monastery in the 13th century. This particular location was chosen for its strategic position near a large salt deposit. Towards the end of the 14th century, Greifswald was involved in the development of the Hanseatic League, and in the 16th century, it achieved the status of a university city — one of the first in north-eastern Europe. Over time, Greifswald developed into an important centre of trade, research activities — particularly in the areas of science and medicine — as well as commerce and administration, and military functions. Thus, Greifswald had always enjoyed a diverse economy, with much of this diversity stemming from its status as a university city as well as its membership in the Hanseatic League.

In the mid 19th century, Greifswald began to experience a first wave of industrial development. In connection to this, several factories for production of machinery as well as a central railway facility were located here. However, this period of industrialization did not have a significant impact on the economy of the city. Furthermore, the closure of the railway plant in the 1920s halted the effects of industry on Greifswald.

It was only towards the end of the 1960s, with the re-location of a communications engineering plant (Nachrichtenelektronik Greifswald, NEG) and the establishment of a nuclear power plant in Greifswald, that industrial activities began to define local economic life. With the establishment of these industries, the subsequent population boom (+ 25,000 people), and the necessary expansion of urban infrastructure, Greifswald began to view itself as an industrial city.

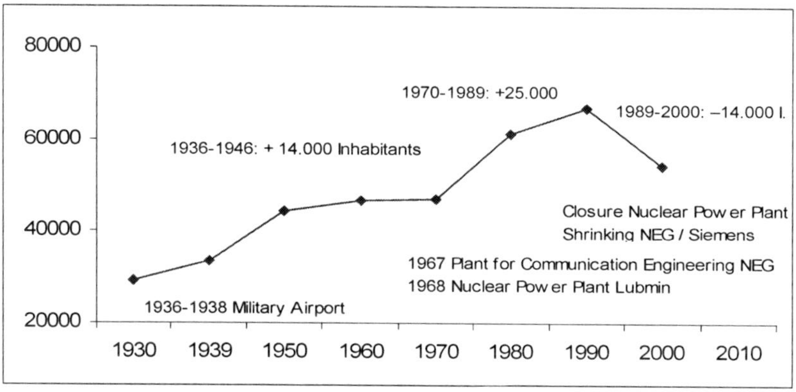

Fig. 1: University and Hanseatic City of Greifswald: Population Development and Defining Events for Urban Development (1930-2000). Data source: Hansestadt Greifswald, Chart created by the author.

Like many other medium-sized cities in eastern Germany, Greifswald has been confronted with serious development challenges since the radical societal and political changes brought about by the end of the GDR in 1989. These problems have been shaped by a complex confluence of socio-economic and demographic changes as well as significant decay of the urban fabric.

Such cities are frequently grouped together under the term "shrinking cities." This term refers to drastic decline of urban population and structure as a result of de-industrialization since 1989, which has led to the closure or downsizing of many industries. Most of these industries were of great importance, both economically and in terms of the status of industrial production in the society of the former GDR, but were no longer viable in the context of the new market economy, or only viable after drastic rationalization and re-structuring.

It was the closure of the atomic energy plant "Bruno Leuschner" and its sub-plants in 1991 that caused the greatest shock related to industrial decline in Greifswald. In 1991 almost all of the seasonal employees as well as 5,000 permanent employees lost their jobs. Only around 1,200 jobs were maintained for the purpose of de-commissioning the plant and will only be maintained until 2007-2008.

In addition, the communications engineering plant (Nachrichtenelektronik Greifswald, NEG) also experienced drastic restructuring. Siemens purchased this company in 1991, and as a result of restructuring the original number of 2,600 employees was reduced to only 360.

Against this backdrop of drastic economic change, it is not surprising that the unemployment rate in Greifswald rose dramatically. Despite the large number of people who left for West Germany immediately after the fall of the Berlin Wall in 1989, the largest changes were seen a few years later as a result of de-industrialization. Between 1991 and 1992 alone the unemployment rate rose from 10% to nearly 17%. After experiencing a slight decrease, the unemployment rate has risen steadily since the mid 1990s, and reached a record high in spring 2005 with 24% unemployment, before falling to 20%, marking the beginning of a period of falling unemployment rates.

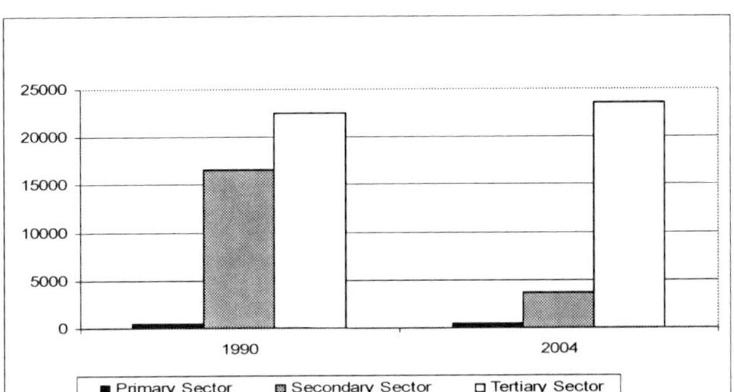

Fig. 2: Employees paying Social Insurance Premiums by Economic Sector. Data Source: Statistisches Landesamt Mecklenburg-Vorpommern, 2006 Edited by the author.

During the same time period, drastic changes were also seen in the employment structure of Greifswald. The total number of employees making contributions to the social insurance system fell by 33% between 1990 and 2005, signalling a decrease of nearly 12,000 employees. While the number of employees in the tertiary sector remained steady — and in fact experienced a slight increase — the total number of employees in the secondary sector fell by nearly 75%. Thus, the distribution of employees amongst the different employment sectors changed dramatically during this period (see figure 2).

Economic decline, particularly in the commercial-industrial sector, coincided with a significant decline in population, caused mainly by the emigration of large numbers of people to more prosperous regions in Germany due to economic reasons. While Greifswald had a population of nearly 70,000 people in 1989, today there are only around 52,000 residents. Thus, Greifswald has lost nearly 25% of its population since the peaceful revolution of 1989.

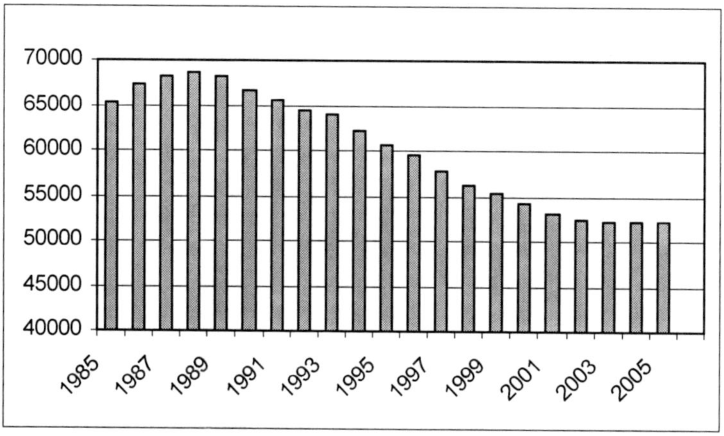

Fig. 3: Population Change in Greifswald (1985-2005). Source: Hansestadt Greifswald, 2002 and 2001-2005. Author's representation.

The first wave of emigration began as early as 1989. During this period, emigration was characterised by people moving large distances to other regions of Germany. After the first wave of emigration resulting from the opening of the border in 1989, emigration in the following years was less marked. However, beginning in the mid-1990s, emigration once again began to increase, beginning a new wave of population decline. In contrast to the first wave of emigration, this period was characterized by a slow and continuous decrease in population, caused by the migration of people from the city to the surrounding region. It was only towards the end of the 1990s that population levels once again began to stabilize.

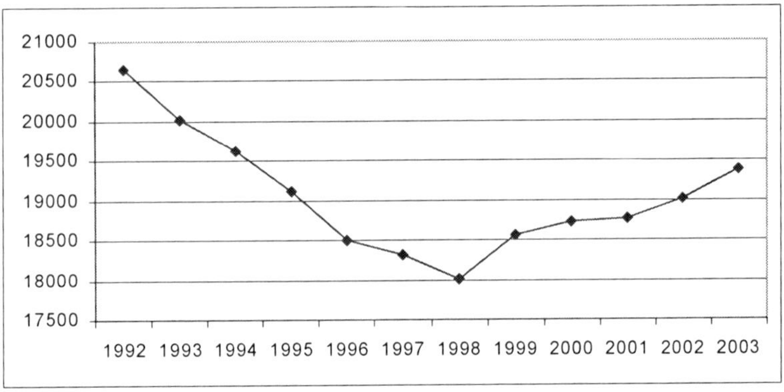

Fig. 4: Population Development (Urban Core) in Greifswald (1992 – 2003). Data source: Hansestadt Greifswald, 2002. Author's representation and calculation.

These changes were accompanied by significant spatial and structural problems, to be observed primarily in the decay of the urban fabric, and were characterized by rapidly increasing vacancy rates in the older areas of the urban core. The successful restoration and refurbishment of older buildings, combined with an increase in the number of vacancies in all neighbourhoods encouraged a slow return of the population to the urban core and adjacent neighbourhoods.

At the same time, the older socialist-era tenement districts, which were often geographically isolated and composed entirely of pre-fabricated medium-rise buildings, experienced a large decline in population that continued throughout the entire 1990s. Thus, these areas tended to experience not only the highest vacancy rates, but with the high visibility of architectural problems, fell out of favour in public opinion, which only served to further strengthen the desire of people to move away from these districts.

Vacancy rates for apartments, and in some cases entire buildings, were particularly low in the newer tenement districts, which had not yet been restored. The problem of population decline in the new districts was complicated by the fact that these areas had not had the time to establish a stable population. Many of the problems experienced by the housing cooperatives and the housing industry in Greifswald were typical of the housing problems experienced in all of the former East German states (cp. Hansestadt Greifswald 2002; Kuder 2005).

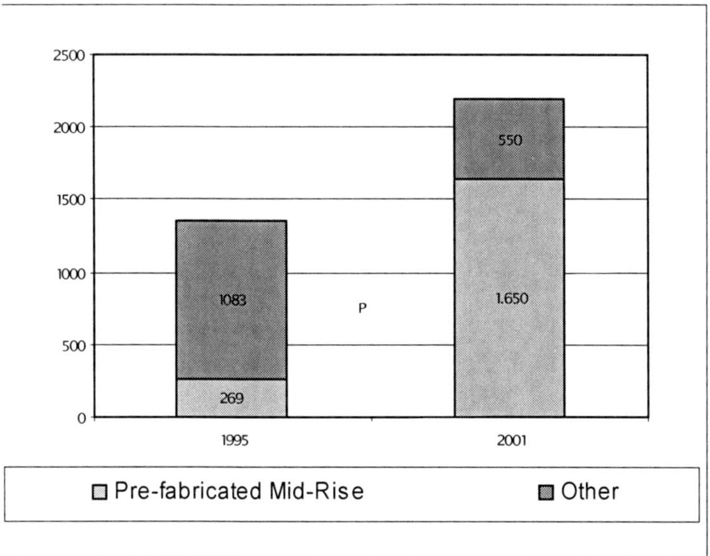

Fig. 5: Vacancy Rates by Building Type in Greifswald (1995 and 2001). Data source: Hansestadt Greifswald, 2002. Author's representation.

4.2 Urban Development Policy in the University and Hanseatic City of Greifswald

After the closure of many of the most important industries in Greifswald, the local economic base was decimated. Surprisingly, in the first few years following these dramatic changes, the attitude of the city remained positive, and was reflected in an urban development policy that was predicated on the expectation that the local economy would soon begin to grow.

For example, at the beginning of the 1990s the city government responded to a perceived need to respond to and "catch-up" with the demands of the market economy through the creation of a "sustainable process of re-structuring and growth" (Hansestadt Greifswald 1992: 9). In particular, it was expected that the local industrial and commercial base could be re-established. As a result, the city administration placed most of its efforts in the creation of new commercial spaces such as technology parks, and attempted to provide an environment in which new technologies could be developed in order to attract new businesses.

The scope of the growth that was expected to result from these efforts is best illustrated through the approach the city government took towards the expansion of commercial space. Thus, existing industrial-commercial spaces of 320 ha were quickly expanded by 245ha of newly created industrial parks (Hansestadt Greifswald 1992: 9).

However, instead of attracting new commercial and industrial businesses, the result was the establishment of a strip of new car dealerships along the perimeter of the city. In a short period of time, around 15 car dealerships set up shop. On the other hand, most of the newly constructed industrial park area remained unoccupied (HGI 01: 41-48; viz. Greifswald Interview 01: line 41-48).

Fig. 6: New "Technologie-Park" Greifswald, 2004. Source: the author.

Commercial activities only began to gain in importance towards the end of the 1990s, after a long period of stagnation. At this time technology-based businesses — particularly in the economic service sector — as well as wholesale and logistics business began to define new kinds of economic activities appearing in Greifswald (Hansestadt Greifswald 2001a: A-16).

Just as significant economic growth had been expected, in the early 1990s it was also expected that the population of the city would grow, bringing with it significant urban development. Thus, in the mid-1990s the preparation of a new land-use plan lead to discussions within the city over its future population numbers, as these projections would serve as the basis for the new plan. Trends in population growth at that time already pointed towards a downward trend, and despite the availability of several pessimistic population forecast studies, the expectation that the city would in fact experience growth formed the basis for approaches towards urban planning taken by the city administration.

The discussion was resolved through official political decision making, whereby it was declared that a population of 70,000 was ideal, and attaining this size became the official objective for all planning activities. Thus, all planning activities, including the creation of the new land-use plan, were now to be oriented towards this number. On the basis of this mythical number, the potential development of the city, the university, and the economy, continued to be seen in a very optimistic light, despite that fact that this level of population growth would never be realized (HGI 03: 31-37).

However, this optimism also had an effect on the comprehensive restoration of the urban core, as well as the further development of the large tenement areas. In particular, the "Ryckzone Development District" — a large area to the north and south of the Ryck River consisting of both greenfield and brownfield sites — was earmarked for extensive urban development projects.

Initially, the underused sections of the Ryckzone between the ports of the Baltic Sea and including the fishing village Wieck, as well as the urban core, were scheduled for development under the banner of "Water-Sports-Leisure". However, this plan was short lived. In 1990-91, development pressure on this area increased sharply, as potential investors suddenly realized its positive qualities (Hansestadt Greifswald 2001a: 12). As a result, a new concept for this area was created, and new development studies for the area were commissioned.

In 1993-94, all urban development studies and reports were consolidated in one overarching urban development master plan. The construction of a modern harbour catering to the needs of leisure watercraft, new housing developments, and the creation of a large housing development with a maritime theme, surrounded by water, and located in the urban core — to be reminiscent of Venice — formed the centre of this plan.

Less than 10 years later, only a few of the planning goals expressed in this document have been implemented. Furthermore, only those goals which are the exclusive domain of the public government have been realized, including urban renewal projects in Wieck, and the expansion of the harbour (Hansestadt Greifswald 2001a: 12).

Other growth-oriented projects were quickly stalled soon after construction began. For example, the development of a new housing development and yacht harbour, based a far distance away from the urban core, was never fully realized. Despite the fact that the area is fully accessible and complete with streets, sidewalks, and lighting, only a few isolated buildings have been completed. Other ambitious projects, such as the maritime-themed housing development or the marina, never made it off the drawing board.

Fig. 7: New buildings in the Ryckzone Development District, Greifswald. Source: the author.

> *"How should I say this? The main wake-up call for the City Marketing was that the city coffers were no longer full enough. The realization that we didn't have enough money to do our job was followed by the question, "why don't we have enough money"?*
> *HGI 05: 214-217*

In summary, urban policy in the first half of the 1990s was characterized by a significant incongruence between the actual state of affairs in the city —characterized by many problems and various crises — and a very optimistic, growth-oriented approach towards urban development. However, despite the many problems that were encountered along the way, it can be said that the city government has remained optimistic in its approach, and this optimism continues to be reflected in its approach to urban development policy today.

> *Together, Greifswald and the surrounding region create a unique entity within northern Germany. This unique entity is a confluence of:*
>
> - *A University City with a long history*
> - *The rural landscape of Western Pomerania*
> - *The coast of the Baltic Sea, stretching between the Islands of Usedom and Ruegen, which provide the ideal backdrop against which to develop tourism, as well as goods, service and technology based businesses with connections to Northern and Eastern Europe*
>
> *The skyline of the old town serves as the undisputable connection between these features. This silhouette is visible from both land and sea. Greifswald enjoys many strong connections to the surrounding region, including unbreakable cultural and economic ties.*
>
> *The University and Hanseatic City of Greifswald and its surrounding region should strive to develop and affirm its status as a lively, innovative, internationally competitive location for scientific research and education that:*
>
> - *Is a desirable place for small and medium-sized enterprises to locate thanks to its accessible technology and service industries*
> - *Enjoys a health-care infrastructure of inter-regional and international importance,*
> - *Possesses an enviable network of innovative and creative educational institutions, at both the secondary and post-secondary level*
> - *Boasts a University with strong ties to industry and technology, that is ecological and multicultural in orientation, and that strives to improve its ties with Northern and Eastern Europe*
> - *Has a high-quality housing, leisure and tourist infrastructure, that is enriched through interconnections with nature, culture, sporting activities, educational institutions, and history*
> - *Enjoys with a full spectrum of services as well as the infrastructure necessary to support a wide variety of meetings, exhibitions and trade shows.*
>
> *The overarching goal is to create and realize an open and flexible urban development policy together with the citizens of Greifswald.*

4.3 Finding a New Leitbild or Overall Concept

Attempts by the city administration to critically reflect upon its previous approach towards urban development policy based on commercial and industrial development, and envision the kind of urban development desired in the future in a way that took the difficult socio-economic and demographic realities into account, occurred relatively late.

Only as the problems associated with urban decline became more serious and more immediately apparent, and as the anticipated economic growth remained elusive, did the necessity of embarking upon a new direction in urban policy become clear. Thus, it was a clear lack of progress in the development of the labour market, despite the many initiatives of the city, that lead to a new, integrated, and interdisciplinary approach towards urban development policy.

In 1996, the city government initiated a project called "City Marketing" as well as another one titled "The Development of a New Overall Concept for Greifswald", begun under the direction of Prof. Dr. H. Klueter, an expert from the Geographical Institute of the Ernst-Moritz-Arndt University in Greifswald. Together with the cooperation of the city administration and an inter-disciplinary project team, an analysis of development potentials and opportunities for urban development that took the economic situation of the city into account was undertaken.

The first step included commissioning a group of experts to conduct an analysis of development opportunities. As part of the process, all departments within city administration were interviewed. The purpose of the interviews was to collect information regarding the different interests and needs of the various departments. In turn, this information was integrated into the creation of an urban development Leitbild or overall concept for local urban planning.

This overall concept for urban development should be viewed as both a formally recognized and socially constructed institution. It is a desirable, widely accepted, and realistic set of clear, normative-anticipatory expressions of urban development goals. Thus, it provides a conceptual framework for the development and creation of more detailed concepts, plans and projects. The overall concept therefore comprises a unified basis for further planning work, and can be seen as a consensus-building instrument regarding the overarching values and goals that motivate the planning process (Durth, Gutschow 1998: 214; Kuder 2004: 56f.).

In a multi-layered participatory process that included policy makers, administrators, parliamentary committees, as well as the public, the resulting overall concept was created by a group of experts and presented as a draft document. This draft document formed the basis for further public discussion. As a result of these discussions, many further modifications to the draft were made. To conclude this informal discussion process, in 1999 the final draft was formally presented to the Greifswald city parliament and adopted by the citizens and city council as an official overall concept, which would provide orientation for the future development policy for the city.

The result of this process was a diverse set of general urban development principles. In addition, the document also identified traditional commercial and industrial opportunities, as well as a broad spectrum of development opportunities. The adoption of this document represented a renewal of the institutional basis for urban development policy, and represented a movement away from the growth-oriented strategies based on commercial and industrial development that had characterized approaches to urban development in the early 1990s.

In terms of the particular strategies, the new overall concept placed a great deal of emphasis on the renewal of the urban core through a new political engagement demonstrated through the creation of new working groups, and thus represented a step away from past approaches based on large construction projects.

As part of the process of creating this overall concept, a new understanding of those factors that represented opportunities for development within the city was encouraged. In particular, the development opportunities presented by the scientific, educational, cultural, and tourism-based sectors of the city were now seen in a new, positive light. Thus, a diverse set of development opportunities was identified, which in turn created the prospect for newer, broader and adaptable possibilities that could be further developed in a variety of ways. However, the breadth of the principles precluded the establishment of a clear and narrowly defined development profile.

At the end of the process, which had taken several years, six permanent city-marketing working groups were formed. Since then, these working groups have carried the responsibility of defining the overall concept in some detail, and have also suggested urban development projects in the spirit of the overall concept. In addition, they have created a mechanism for monitoring and evaluating the projects that have resulted from the overall concept (Hansestadt Greifswald 2006).

Following the advice of the external consultants, the thematic organization of these working groups represents a departure from the traditional structures of the city administration. The following working groups, formed during the initial process, continue to exist today: A) Medium-sized Enterprises, Technology, Service Sector, and the Economy; B) Health as a Location Factor; C) Education; D) Housing, Recreation, Culture, Sports, Youth, Nature and Tourism; F) The Greater Region and G) The Revitalisation of the urban core.

In contrast to the process through which the development opportunities were identified and the overall concept developed – which was largely led by experts with occasional opportunities for broader participation "in the shadow of the hierarchy" (Scharpf 1973) – the realization and implementation of the concept was part of a long and complex governance process, involving a wide range of very active governmental and non-governmental actors, and lead by the working groups.

These working groups included a total of 90 people and were thematically organized. These groups suggested more than 200 small-scale interventions as well as several larger projects related to urban development (Hansestadt Greifswald 2006).

These projects were in turn implemented through the participation of representatives of various administrative areas who were active members of the working groups and who were able to bring the ideas forward to the city administration. The working groups were then able to monitor the implementation and realization of these projects.

Although the overall concept, as it was originally defined, was only a vague sketch of a desired future, the process of developing and realizing certain ideas through the working groups lead to a deeper understanding of the concept and consequently to its sustainable implementation. Consequently, a better understanding lead to both greater public and political commitment towards and support for the concept, as well as the political and administrative institutionalization and integration of the public working groups within the political- administrative apparatus.

The significance of this overall concept for Greifswald can be measured using the example of the Ernst-Moritz-Arndt University. During the process of developing and implementing the overall concept, the university was increasingly viewed as a central factor for urban development for Greifswald. As a result, the development of the university became a focal point for urban development policy. As a newspaper article, published towards the end of the 1990s in the *Frankfurter Allgemeine Zeitung* (1999), declared in a headline: "In Greifswald, the University Has Returned to Centre Stage."

> *Anyone who slanders the university also slanders the whole town. Of course, they might see it otherwise in Berlin or Rostock. But when we continue to cut (the university) then we can also continue to tear down houses. HGI 02: 111-114*
> *Without the university, the city is dead. It's just that simple. HGI 04: 408*

Considering the goals expressed in the overall concept, the city government adopted this article title as a slogan, and in 2002 signed a cooperation agreement with the university. The agreement set the stage for a new era of cooperation between the city and the university (Hansestadt Greifswald 2003: X, XI). In the same year, the city administration began the process of creating a Tourism Plan for Greifswald with the cooperation of the Geography Department of the university. In the following year, the plan was completed and became the official policy on urban and tourism development.

Finally, the spirit of cooperation between the university and the city was demonstrated in 2003-04 when Greifswald, in light of the new recognition of the importance of the university, submitted a formal application to receive national recognition as "Science City 2005". This application was a joint effort between the city and the university.

Although the application was unsuccessful, the decision was nevertheless made to officially rename the city as the "University and Hanseatic City of Greifswald" (*Universitäts- und Hansestadt Greifswald*). With this re-naming, the city sent a very strong message to the outside world regarding the importance of the university within the city. In addition, the re-naming meant that the university became a very important part of the official image of the city, and cemented the role of the university in future development plans (Hansestadt Greifswald 2003: 15).

The overall concept continues to provide a basis for urban development today. This is partially due to the fact that the realization of this concept has been enabled by a variety of support mechanisms. After the successful implementation of many initial projects, the working groups have been able to adapt and develop both their content and format in order to meet changing situations. As a result, over time, the working groups have changed in both their structure and orientation.

For example, one working group, concerned with improving the overall quality of life for residents of the city, has been discontinued because it was perceived that the interventions necessary to improve the quality of life would be undertaken by other working groups. Another working group that had concentrated primarily on construction projects and physical improvements that could contribute to the revitalization of the urban core has drastically reduced its activities after achieving many of its initial goals. On the other hand, the working group that focused on the situation of Greifswald in the context of the greater region has become increasingly active due to administrative boundary reforms in the State of Mecklenburg-Vorpommern.

Another working group, which focused on fostering the economic success of the health sector, has developed into an independent organization, and is now operated on a professional basis. This professionalization was in direct response to the considerable demands posed by the complexity of the health sector. However, the activities of this organization continue to be guided by the aims of the overall concept.

To characterize the effects the creation of the overall concept has had on the approach taken towards urban development today, it must be said that the city administration has identified a wide variety of development opportunities and created a diverse set of urban development initiatives, particularly in the areas of commerce and industry. Furthermore, those areas that are connected to the university, such as research and development, particularly in the health-related fields, as well as cultural and tourism development have also been the focus of much attention. Thus, the city administration has attempted many different initiatives in these areas, despite continuing difficulties particularly in the labour market.

In addition, the city-marketing working groups have been able to establish themselves as an important institution within the city. These groups have been able to realize and further develop the initial overall concept through further projects and initiatives. These development initiatives have enjoyed broad support. In particular the more promising initiatives in the areas of medical technology have been supported to reach their full potential.

The process of creating an overall concept for urban development in Greifswald is distinguished in how it both opens the way for the discovery of new and diverse paths towards urban development, without hastily closing other, less obvious approaches. In addition, meaningful support is provided for the opportunities identified.

4.4 Analysis

An analysis of changes in the approach towards urban development in the University and Hanseatic City of Greifswald, and a discussion of the accompanying institutional changes, will highlight the ways in which the process of creating an overall concept has affected these changes.

The significance of the Leitbild or overall concept for development policy in Greifswald can be particularly well demonstrated through an analysis of how they influenced institutional change. Institutionally, Greifswald had been very stable up until 1998. The creation of a new overall concept represents an attempt to deal with the sudden and drastic change in the local political and economic context, and the many problems that accompanied these changes. The creation of an adequate basis for an institutional re-orientation was enabled in part through the creation of new approaches and strategies, as developed by the city marketing groups, and which in turn provided the basis for the further development of the overall concept.

The process of changing direction in urban development policy, as discussed in this case study, was thus not a revolutionary change, but was rather the result of a complex set of changes over time. The beginning of this process can be traced back to a perceived moment of crisis in light of the many social and structural changes that the city experienced in a short period of time, particularly when it faced economic and demographic decline and recognized the serious financial crisis it was confronted with. High unemployment rates, the steady decline in population, and failures of the housing market, particularly in the soviet-style pre-fabricated housing stock on the edge of the city were issues of particular concern. Problems in the local housing market were viewed as an indicator of the failing local economy, and housing market interests demanded a strategy that addressed this specific problem.

As a result, Greifswald experienced destabilization and period of uncertainty. Although this period of uncertainty did not immediately follow the crisis, it nevertheless led to a questioning of traditional approaches towards urban development policy. For example, players in the local housing market demanded that plans for new housing developments that had been planned with a view to promoting economic growth be shelved, in order to lessen competition on the housing market.

The potential of traditional urban policy to achieve its stated goals was thus placed under greater scrutiny. In addition, the political desire to modernise and replace older institutions and mechanisms in order to better respond to the new realities of the market economy also signalled a turning point in Greifswald's institutional development.

On a theoretical level in regard to Leitbilder or overall concepts, the recognition of institutional uncertainty and need to modernise supports the thesis that new principles always become important at the point when the usual assemblage of norms and orienting markers becomes destabilised and need to be replaced by a new navigational framework (Streich 1988: 135).

Consequently, this turning point was shaped to a large extent by the process of constructing an overall concept for the future within the framework of urban marketing. This process was intentionally disengaged from the traditional sectoral divisions within the city administration, and instead, lead by external experts. Thus, new directions for urban development policy could be developed independently of the traditional structures of the city administration. In turn, new ideas for urban development policy could be discussed and integrated within a clear and thematically organized, inter-sectoral overall concept.

The draft overall concept was then further developed and finally adopted as a legally binding document as the official Leitbild of the city government. The draft document had however played a very important role, and fulfilled a very large set of responsibilities in leading up to its official recognition (Kuder 2004). The draft document thus prepared the way by demonstrating both the possible range of development potentials as well as a number of ways of expressing these potentials, and also served to mobilise city stakeholders in discussions regarding the future of the city. This discussion formed an important governance exercise, and also increased overall support for the official document when it was completed.

A further element which defined the turning point for urban development policy in Greifswald were the ways in which the concept received further development and redefining through the city marketing working groups. It was only in the context of these groups, in which concrete projects were proposed and developed, and through the development of over 200 realistic projects and small interventions, that the overall concept began to be better understood. While the overall concept had originally been expressed in vague terms, it was through this iterative process including adequate institutional support that it could be transformed into a consistent, comprehensive tool for guiding urban development.

This process was thus able to overcome the problem of bridging the gap between policies developed in relation to an abstract overall concept and the creation of projects that would be realistic and workable on the ground. In this context, the overall concept also served a greater function, independent from the process of developing urban policy. The concept therefore also orients the derivation of further goals, and provides a set of indicators for monitoring the success of projects implemented.

Thus, not only the overall concept itself, but also the process of developing it had a major impact on local urban development as well as on the institutional structures and contributed to institutional change within the city administration. The process of developing this concept therefore highlights the extent to which the so-called "turning point" was actually part of a complex set of processes including the questioning and dismantling of old methods as well as the development of new practices. This process is in line with cyclic theories of innovation, such as Thomas S. Kuhn's (1973) notion of the paradigm shift, which encompasses the cyclic notion of questioning and dissolving old paradigms, and the subsequent development of new paradigms.

Finally, the fact that the concept lead to a diverse set of directions and initiatives in relation to urban development policy has already been mentioned, and will now be discussed in further detail.

The central function of the overall concept was not to provide a strict set of fixed ideas for development, but rather to provide a framework from within which further ideas could be developed, while at the same time reducing a large amount of the complexity that arises if all desired and possible paths are considered when envisioning desirable future outcomes. This concept is thus broad enough to allow creative approaches, while at the same time providing a logically consistent set of parameters for development.

However, in no way does the overall concept constitute a fixed set of policies, nor does it comprise a vision for the future that is unchangeable or inflexible. Nevertheless, they do play an important role in creating a foundation for the political culture of local urban planning.

At the same time, several theoretical approaches will suggest that the process of creating an overall concept or a period of institutional change will not necessarily result in the formulation of a clearly articulated goal for the future. For example, institution theorists will argue that the construction of new institutional pathways can come about as the result of a variety of new and parallel approaches.

Although these approaches may appear to be in conflict with one another, positive feedback, in the form of increasing returns and moments of success (e.g. the creation of new business), will eventually lead to a decision in favour of one path of development.

Friedrich's (1993) theory of urban decline provides a way to better understand this phenomenon. This theory, which approaches urban decline from the perspective of product cycles, suggests that the differentiation of the economy, for example through the founding of new economic enterprises from a variety of sectors as seen in Greifswald as a result of the new overall concept, in itself presents a significant goal for a new overall concept. Diversification can thus be understood as a central element of a clearly defined and goal-oriented urban development policy.

Cycle and paradigm oriented theories offer an alternate interpretation of the life-span of overall concepts. Thus, concepts are understood to be in a constant state of development, refinement, and selection. This process is defined by the continuous presentation and negotiation of competing interests and ideas (e.g. Kuhn 1973).

Only when an overall concept has become crystallized through this process and has taken the form of a more highly developed set of ideas regarding future urban development, and some degree of success has been experienced, it can be understood as a truly mature and developed overall concept. It is at that stage, when it has received recognition from both, the broader public and planners alike, and becomes, in Thomas S. Kuhn's formulation, a "promise of success," that an overall concept can be understood as a development paradigm.

In any case, regardless of how one decides to interpret this case, the importance of the development of an overall concept — whether or not it has been fully developed into a paradigm or not — has played a very important role in the creation of urban development policy in the University and Hanseatic City of Greifswald.

In addition, the case study demonstrates the importance of implementing new, diverse, and creative approaches to urban development, through the initiation of a creative process not entirely free of conflict that involves all possible actors and stakeholders, with the goal of developing a Leitbild or overall concept. Especially in instances where cities have experienced rapid change and crisis, or where the tried and true methods, approaches, and values have proven to be unsuccessful, it is imperative to follow a principle-oriented approach with patience and determination, if any success for future development is to be realistically expected.

References

Bürkner H.-J., Kuder T., Kühn M. (2005): Regenerierung schrumpfender Städte. Theoretische Zugänge und Forschungsperspektiven. Working paper. Erkner.

Durth W., Gutschow N. (1988): Träume in Trümmern. Stadtplanung 1940 -1950. München.

Frankfurter Allgemeine Zeitung (1999): In Greifswald steht die Universität wieder ganz im Mittelpunkt. 23.02.1999, No. 45, 17

Friedrichs J. (1993): A Theory of Urban Decline: Economy, Demography and Political Elites. In: Urban Studies, Vol. 30, No. 6, 907-917.

Hansestadt Greifswald (ed.) (1992): Gesamtstädtischer Rahmenplan. Diskussionsergebnisse zur Gliederung und Entwicklung der Hansestadt Greifswald.

Hansestadt Greifswald (ed.) (2001-2005): Statistische Vierteljahreszahlen. http://www.greifswald.de.

Hansestadt Greifswald (ed.) (2001a): Wettbewerb Stadtumbau Ost. Integriertes Stadtentwicklungskonzept.

Hansestadt Greifswald (ed.) (2002): Statistische Jahreszahlen 2001.

Hansestadt Greifswald (ed.) (2003): Leben Lernen Wissen. Greifswald im Wettbewerb Stadt der Wissenschaft 2005. Application brochure.

Hansestadt Greifswald (ed.) (2004): Sozialanalyse 2004.

Hansestadt Greifswald (ed.) (2006): Internet-Portal.

Kuder T. (2004): Nicht ohne. Leitbilder in Städtebau und Planung. Von der Funktionstrennung zur Nutzungsmischung. Berlin.

Kuder T. (2005): Problemgebiete ostdeutscher Stadtentwicklung - Beobachtung aus der Begleitung des Städtebaulichen Denkmalschutzes. In: Kirchhof and Franke (eds.), Städte im Umbruch, No. 3.

Kuhn T. (1973): Die Struktur der wissenschaftlichen Revolution. Frankfurt/Main. Scharpf F. W. (1973): Planung als politischer Prozeß. In: Scharpf F.W.: Planung als politischer Prozeß, Aufsätze zur Theorie der planenden Demokratie. Frankfurt am Main.

Streich B. (1988): Grundzüge einer städtebaulichen Leitbildtheorie. Bonn.

Statistisches Landesamt Mecklenburg-Vorpommern (1991-2006): Statistisches Jahrbuch Mecklenburg-Vorpommern. Schwerin.

5 Growth and Innovation through Urban-Rural Partnership

Rupert Kawka

Introduction

In June 2006 the Federal Ministry of Transport, Building and Urban Affairs and the State Ministers responsible for regional planning passed the new guiding principles for spatial development in Germany – entitled "Growth and innovation", "Ensuring services of public interest" and "Conservation of resources, shaping of cultural landscapes". Whereas latter two concepts were widely accepted, the subsequent discussion concentrated on the first one. There was consent that growth and innovation are important, however also dissent on the question which regions should generate growth and innovation. Many critics regarded the concept as neoliberal for several reasons. Firstly, they presumed that spatial planning regards merely the metropolitan regions as the most qualified ones to generate economic growth and innovation. Thus, they accused spatial planning of considering rural areas as residual regions. Secondly, many misread the guiding principles: It was said that the first concept intends strengthening of the strong regions – i.e. the metropolitan ones – instead of increasing the strengths of all regions, as the text states (Standing Conference of Ministers responsible for Spatial Planning 2006: 12). This created the fear that the guiding principles intend to redirect subsidies from economically weak to strong regions, i.e. to strengthen a growth policy at the expense of a balancing policy. But these aspects do not apply – the criticisms are based on either unconscious misunderstandings or deliberate misinterpretations.

Presently, spatial planning is in a state of bringing the guiding principles into action. But for a successful implementation and public acceptance, the above mentioned fears have to be taken seriously. Thus, the basic question arises: Where does this renunciative position stem from?

5.1 A Short Glance at the German Situation

Balancing policy has a long tradition in Germany. The German constitution is assigned to provide for equivalent living conditions all over Germany, and the strong lobby for rural areas – e.g. the farmers' association (*Deutscher Bauernverband*) and the association of the counties (*Deutscher Landkreistag*) – which are one the strongest critics of the first guiding principle, take this passage from the constitution as the basis of their policy (cf. Nova-Institut 2006: 6-7). But also other programmes with heavy subsidies, e.g. the promotion program for the border regions to the GDR and Czechoslovakia until 1990, show the strong focus on balancing policies rather than on growth policies.

The first guiding principle has been a trial to strengthen growth policy – but not at the expense of the other areas. Actually, it tried to combine both aspects – a growth policy and a balancing policy, i.e. to care for growth and innovation, but also for regional balance between different types of regions – urban and rural, central and peripheral, as well as strong and weak ones, as will be shown in the following chapters.

And looking at the German situation, there is the necessity to have a closer look than before at both the central and also economically strong regions, so that they remain strong and provide means for balancing policies. A simple empirical observation shows how the regions contribute to the national growth of GDP, i.e. which regions are the growth engines of the national economy. The answer is given in fig. 1, and this graph is also one main empirical foundation of the concept titled "Growth and Innovation".

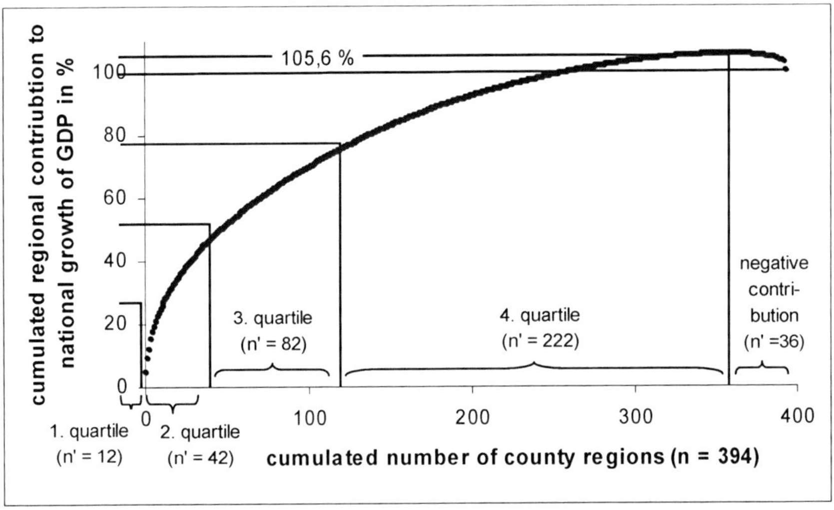

Fig. 1: Contribution of regions to growth of national GDP. Sources: www.vgrdl.de, author's calculation.

The empirical observation – of course, no theoretically based law – can be called „15/50-rule“: In the period of 1994 to 2003, the strongest 15 % of the German regions - altogether 54 out of 395 county regions – generated 50 % of economic growth, whereas the weakest 50 % of the regions contributed only 15 % of the national growth of GDP. The major contributors are the large cities with their surroundings counties, but this is not automatically the case, e.g. Berlin and Hanover do not belong to this group of regions. Contrary, also some rural or urbanized areas, even a few peripheral ones, are within the group of the strongest 15 % of the regions (Kawka 2007: 39). Examples are Bodenseekreis, Emsland, or Dingolfing-Landau. This is a hint that a simple division between "urban = economically strong" and "rural = economically lagging" is neither adequate nor empirically grounded and that a growth policy cannot be simply a metropolitan policy. But it is also a hint that the larger cities and their surrounding counties have an outstanding importance for the whole country and especially for the regions benefiting from balancing policies.

Furthermore, due to the polycentric settlement structure in Germany, this kind of policy would not be beneficial for merely a few cities, as there are not one or a few dominating centres in Germany, but many larger cities across the entire territory, except for the north-eastern regions. The distribution of big cities is less important in this respect. Of higher interest is the distribution of metropolitan functions, i.e. of functions normally located in larger cities, such as airports, trade fairs, the seats of political decision makers etc. (Bundesamt für Bauwesen und Raumordnung 2005: 178). Fig. 2 shows the distribution of metropolitan functions in Germany, indicating their polycentric pattern. Strengthening the metropolitan functions as a location factor for economic development is no policy merely for the urban areas, but also for the benefit of the regions surrounding these larger cities, i.e. also for rural and urbanized places, as these metropolitan functions have a large catchment area.

Of course, there are policy programmes promoting universities or research and development, and the places with research and teaching institutions benefit from these policies. But these are sectoral policies which are not integrated in an explicit spatial policy for growth areas.

The two examples show that metropolitan regions have a high economic and functional importance. Spatial planning has to take this outstanding position into account. But a growth policy can only have a realistic chance to be accepted when it also includes an option for rural areas. Due to the polycentric settlement structure and the existence of strong rural areas (symbolized with stars or suns in fig. 3), urban-rural partnerships can be regarded as the appropriate approach in this respect. This is the basic idea of the first guiding principle which is also an option for the areas strongly lagging behind, i.e. the regions with a need for stabilization, where unemployment and out-migration is high and economic growth is low. The strategic element is to combine the types of regions with their specific potentials and also their specific needs and to form large-scale networks for responsibility (*"großräumige Verantwortungsgemeinschaften"*). The main aim of this approach is to achieve growth and innovation as well as cohesion and balance on a regional

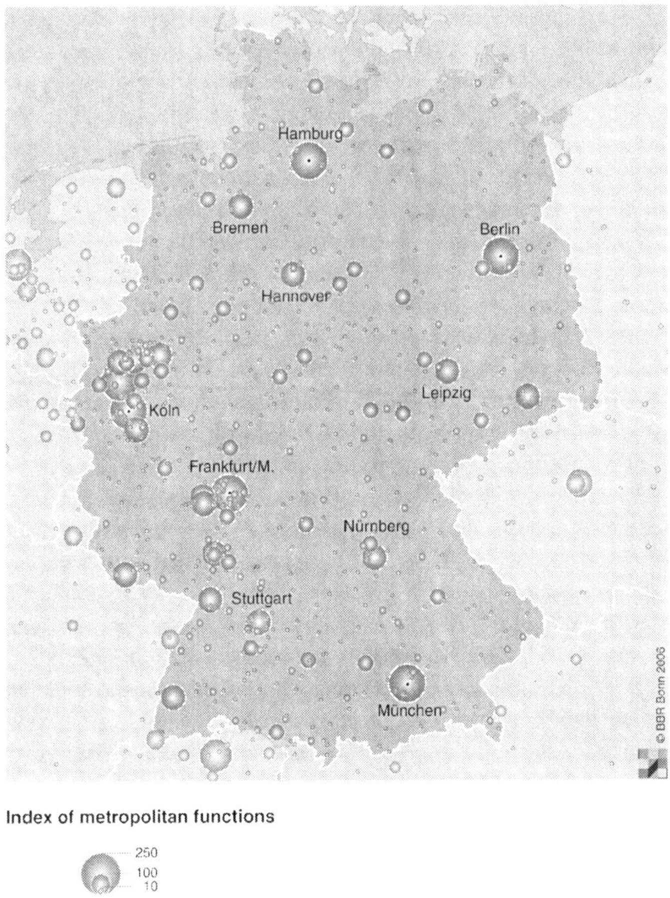

Fig. 2: Distribution of metropolitan functions in Germany.

level between these different types of regions: Central and peripheral, urban and rural, economically strong and weak regions should contribute to growth and innovation according to their potentials. But all regions within their larger spatial context should have the chance to participate in growth and innovation generated within this spatial context. Mainly due to this element, this first guiding principle cannot be regarded as neoliberal and focussing merely on the metropolitan areas.

The strategic approach of urban-rural partnership is not really new in Europe. About a decade ago, it came into being mainly through the European Spatial Development Perspective (ESDP). Today, urban-rural partnerships are en vogue, because co-operation always sounds good. These kinds of partnerships apparently bring together the most contrasting types of regions and contribute to solving the antagonism between town and countryside. And they even use these regional differences to form a unity – like yin and yang – and a surplus value. Thus,

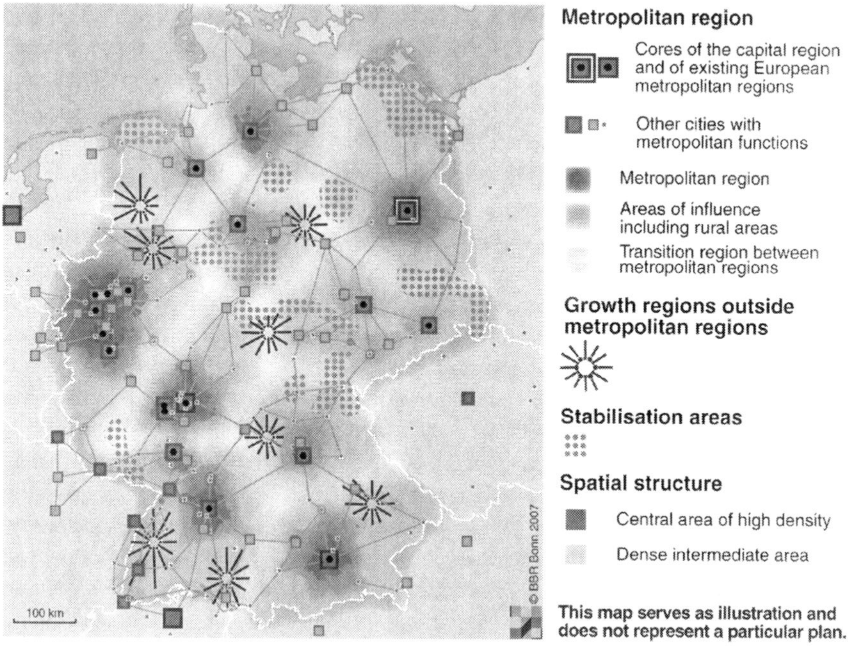

Fig. 3: Concept "Growth and Innovation".

politicians and spatial planners seem to have found the magic formula to overcome spatial disparities. Urban-rural partnership is easily said, and the concept suggests win-win situations.

Indeed, the idea is good, but this instrument has not matured yet. Thus, criticism and cautious remarks about the feasibility of urban-rural partnerships have to be taken seriously. Although a number of examples for urban-rural partnerships exist in Europe, there are many open questions about this idea, e.g. to which extent these partnerships contribute to cohesion, if they can stimulate economic growth, which regional factors promote or restrict them, and which governance models between the different types of regions are appropriate. Thus, the proof is still lacking that urban-rural partnerships really work, but without trying them, the answer will be never achieved.

Despite these partly ironic observations, urban-rural partnerships are indeed a kind of balancing policy on the regional level. Their goal is to give as many people as possible, irrespective of their places of living – either in urban, suburban or rural areas – the chance to participate in spatial and especially in economic development. Furthermore, they are regarded as an important element for supplementing the polycentric urban structure by improved integration also of rural or peripheral areas. Urban-rural partnerships are not easily implemented, and they do not automatically result in success stories in every region. It is also possible to find many arguments and examples showing that they hardly work. Looking at experiences in other countries – as done in the following chapter – may comprise the basis for a further reflection on this topic.

5.2 General Remarks on Urban-Rural Partnerships

The following paragraphs provide no theoretical reflection on urban-rural partnerships, because a theory on this topic does not exist. The paragraphs are not really an empirical observation either, as no evaluation of urban-rural partnerships exists yet. The following observations could be rather regarded as commentary on a topic gaining more and more importance. And a view of some examples can provide a better understanding of the chances and limits of this tool and the implications for this policy in Germany.

The topic of urban-rural partnerships is a vital one in the EU and in many European countries. This is mainly due to the ESDP of 1999 (European Commission 1999), as this document is one of the first important political sources mentioning the topic. P. 11 of this document provides the following quotation: "As early as 1994, the Ministers responsible for spatial planning agreed on three policy guidelines for the spatial development of the EU: development of a balanced and polycentric urban system and a new urban-rural relationship; securing parity of access to infrastructure and knowledge; and sustainable development, prudent management and protection of nature and cultural heritage." Eight years after passing the ESDP, the idea is still vital, even if the EU is not yet completely covered by urban-rural partnerships. The Territorial Agenda of 2007 repeats and even stresses the goal – „we need new forms of partnership and territorial governance between rural and urban areas" (Informal Ministerial Meeting on Urban Development and Territorial Cohesion 2007: 5). Thus, these documents provide a sound political frame so that efforts on a subordinate spatial level can implement this tool.

Nevertheless, the reactions to these two documents are divergent, i.e. the Study Programme on European Spatial Planning (Bundesamt für Bauwesen und Raumordnung 2001) based on the ESDP has provided many ideas on this topic. These are still up to date; and on p. 146f. the Study Programme also mentions some 38 initiatives as case studies for urban-rural partnerships. This, in principle, indicates the interest of regions in urban-rural partnerships. But there are also a number of regional political documents or development plans simply mentioning the term "urban-rural partnership". This is often simply name-dropping to pay homage to the ESDP and the Territorial Agenda – without trying to perform it in practice. The studies of Scott Wilson Business Consultancy (2006: 2) or Comhairle Nan Eilean Siar (2002: 35)just serve as two examples for such lip service.

Furthermore, there are a number of documents and initiatives confusing urban-rural partnership and urban-rural relationship. A relationship is the precondition for a partnership or, at least, a good starting point to form a partnership, although, of course, a clear cut distinction between both is hardy possible, as there is rather a smooth transition between the two terms. But not every commuter flow or every provision of potable water, wind energy or food products is a partnership. An example for such a misunderstanding is the publication of Polverani, Quiogue, Gross and Novotny (2005). On p. 45, the authors list three cases – Tuscany in Italy, Niederösterreich in Austria, and Norra in Sweden – as examples. But only the Austrian one has a connection to urban-rural partnership, as the Italian case deals with urban regeneration (e.g. revitalisation of urban market centres) and the Swedish case with endogenous potentials in rural regions without large urban agglomerations. Even if rural dwellers benefit from an urban development project, it is neither an urban-rural relationship nor an urban-rural partnership – and readers can get the impression that urban-rural partnerships are insubstantial entities. A further example is the paper of the Council of European Municipalities and Regions (2003). Although the introduction focuses on co-operation between urban and rural areas, the following case studies hardly focus either on urban-rural partnership or on urban-rural relationship. Out of nine case studies, only two refer to the main topic of the introduction, and both of them describe processes in the same region. The majority is however only devoted to topics in single communities. Thus, there is a high discrepancy between the introduction and the following case studies of that volume, showing a general misunderstanding of the topic. The involvement of a governance model makes the difference between a partnership and a relationship.

The examples show that urban-rural partnerships are no easy topic, and their creation is a challenge for the regions and the regional actors. Zimmermann (2007: 214) is also sceptical that areas of corporate responsibility (*"Großräumige Verantwortungsgemeinschaften"*) will develop – in his eyes at most in the urban-suburban context, because functional relationships exist mainly between closely neighbouring areas. Distance between the actors and regions may be an obstacle for cooperation. Indeed, this influence seems to be underestimated, such as e.g. in the project "Berlin and the complementary region Usedom" of Baltic+ (Baltic+ 2004: 14-17). The 200 km distance between these two regions may still allow a relationship, but it is hard to imagine that a real partnership can arise. Distance, and with distance the increasing difficulty for functional relationships, for co-operation, and for mutual understanding and responsibility is very crucial in this respect. And distance is also a major difference in the understanding of urban-rural partnerships all over Europe. As shown in the introduction and also in the next chapter, the German understanding is based on a broad scale incorporating different types of regions. The interpretation e.g. of the Romanian Government is in contradiction to this view. Under the headline of urban-rural partnership, it wants to encourage "the cooperation [...] especially between close localities [and the i]dentification of the cooperation possibilities between close urban and rural localities" (Government of Romania 2005). And a third position is provided by the Swiss understanding: According to the opinion of the expert commission for reviewing the new concept of regional policy in Switzerland, urban-rural partnerships have to promote innovation and value added in rural regions. In their eyes, it is irrelevant where the centres are spatially located – even in remote foreign places. And they give the example for a co-operation between a Swiss rural region and an American technological cluster (Expertenkommission "Überprüfung und Neukonzeption der Regionalpolitik" 2003).

The broad variety of spatial definitions may be as necessary as the regional setting and the goal influencing them. And the differences vary between countries and also within them, e.g. in Hungary, the OECD (2001: 146) mentions: "During the socialist era village-town relationships were forced to be highly subordinated, so local village authorities often refused to co-operate with towns. Villages usually establish micro-regional development associations with no participation from bigger towns. This serious lack of urban-rural partnership further hinders efficient planning and development at the local, micro-regional level." In contrast to this, there has been a broad variety of projects for promoting co-operations between communities, e.g. via urban networking or urban-suburban collaboration. So, historical experiences can play an important role in bringing urban-rural partnerships as the most complex form of co-operation into action easier in one country than in another one.

Furthermore, rural regions form no homogeneous group (Kawka 2007: 104). The variety is as large as in the case of urban areas. Some are, e.g., economically stagnating and losing their inhabitants due to outmigration; some are economically prosperous with a positive net migration. Urban-rural partnerships have to take the specific strengths and weaknesses of the partner regions into account, and they determine the topics for the cooperation; e.g. networking between rural and urban companies is more appropriate between economically strong regions, but an urban-rural partnership based on tourism can be a basis for regions with a beautiful landscape and potentials for leisure activities. The improvement of public infrastructure can be a topic for a partnership between weak regions with many commuters to urban places with a good economic performance. Thus, experience from one region cannot be easily transferred to another region – but, of course, to some extent. This is even truer for international findings. One country cannot simply copy the examples from another country, as e.g. in addition to the level of intermunicipal and regional co-operation, also cultural factors have an influence. For instance, the culture of having second homes for weekends is much more widespread in France or Scandinavia (Steinecke 2007) and less so in Germany, so that the option to build rural-urban relation- and partnerships based on this issue will have different chances for success in different countries.

Nevertheless, the number of initiatives – however defined – has increased during the last years to bring urban-rural partnership into action. These initiatives have provided good examples showing that urban-rural partnerships can work. Furthermore, the broad variety of topics demonstrates that there is not only one good solution but several. They range from regional chains of value added, e.g. in foodstuff – examples are "Regionalmarke Eifel" (cf. www.regionalmarke-eifel.de) in the region between Köln, Koblenz, Trier, and Aachen or "Unser Land" (cf. www.unser-land.info) consisting of nine regions around Munich in Germany – to infrastructure projects (the eastern Norway county network around Oslo) as mentioned by Metropolitan Areas+ in the Baltic Sea Region (Metropolitan Areas Baltic Sea Region 2006), or the joint regional planning of Berlin and Brandenburg (Senatsverwaltung für Stadtentwicklung, Ministerium für Landwirtschaft, Umweltschutz und Raumordnung, Gemeinsame Landesplanung Berlin-Brandenburg 2007). And it can be expected that the German demonstration projects, described in chapter 4, will provide more good examples for urban-rural partnerships. Also large European metropolitan regions which include different types of regions can be regarded as a form of urban-rural co-operation which even include a governance model.

Many of these good examples are related to economic topics, and they often take the potentials of rural regions into account. Nevertheless, rural areas are often regarded – or they regard themselves – as the inferior partner in urban-rural partnerships, as the potentials of the cities are higher. On the other hand, urban areas are considered dominating or, at least, trying to dominate the partnership. Of course, cities provide, as a general rule, more employment than rural areas, and the commuter flows to the cities prove this to be correct: e.g., productivity is mostly higher in urban than in rural areas, and infrastructural amenities as well as shopping facilities are better there. Nevertheless, rural areas also have their strengths and complimentary potential. Open space, landscape, quietness and clean air, just to mention a few, are important values and public goods free of charge. Additionally, rural areas are important for the production of food, the preservation of ground water, for residential purposes, and thus for human capital living there and working in cities. Furthermore, also important economic centres are located in rural regions. Examples are the rural growth centres mentioned in the first guiding principle (fig. 3). Nevertheless, there is still an asymmetric relation between urban and rural areas, due to the quantity and the often the quality of these potentials (c.f. Bengs and Zonneveld 2003: 285). The Territorial Cohesion Report mentions this obstacle on p. 27f, too: "Although rural areas play a vital role in everyday life and in the modern economy, these areas are in many ways dependent upon economic activities and facilities located in cities and urban areas. Although some crude forms of domination from the urban areas have disappeared, other softer forms of domination have emerged, such as the transformation of rural areas into consumption landscape" (DG Regional Policy 2004). But a partnership is only possible when all actors are on the same eye level. Thus, the finding of an adequate governance model is necessary for successful partnerships, so that the regions can use their specific potentials within the larger functional region.

5.3 Supraregional Partnership – Experience from Germany

General Remarks on Demonstration Projects

In Germany, the federal level of spatial planning implements the strategic approach of the new guiding principles as mentioned in the first chapter, i.e. the formation of urban-rural partnerships. But the federal level has only limited means concerning spatial planning. Therefore, demonstration projects ("*Modellvorhaben der Raumordnung*" or, in short, MORO) are used since 1996 (Gatzweiler 2006: 6, 7, 9). MORO is an informal planning instrument and has proven to be quite powerful for developing good or best practices for different topics related to spatial planning. Throughout the last twelve years, over 500 subprojects on different topics have been supported. The financial resources come from the Federal Ministry of Transport, Building and Urban Affairs, and the Federal Office for Building and Regional Planning puts MORO into execution. These projects serve as examples for good or best practice showing other regions ways to cope with spatial aspects. The MORO-budget is merely about 1 Mio. Euro per year. This money is not enough for investment, but it helps the regions to organize the processes and to cover the costs, e.g. for a scientific monitoring. Despite this comparatively small amount of money, experience of the past has shown that many regions want to become demonstration regions and to try new approaches in regional development.

Demonstration Project "Supraregional partnership"

Co-operations between communities have always been a major topic for MORO, e.g. urban networks or co-operations between cities and their nearby hinterland have been topics of the last years. Thus, the new demonstration project "supraregional partnership" is a continuation of this tradition.[1]

At the end of June 2007, the demonstration project "Supraregional partnership, innovative projects for urban-regional co-operation, networking and shared responsibility" was launched. As knowledge of already existing or planned initiatives on urban-rural partnerships in the regions was limited, the project started with a call of interest. Some 60 initiatives showed their interest in the demonstration project. Not all initiatives matched the expectations of the original idea, but a short list containing some 15 initiatives was compiled. They were invited to a workshop in Bonn in September 2007 to present their project ideas and to discuss them. Out of these 15 initiatives, seven were selected for a financial promotion.

Fig. 3 shows the location and spatial extent of the demonstration regions, and the following list gives a more detailed description of them from north to south:

The partnership in northern Germany (*Großräumige Partnerschaft Norddeutschland/Metropolregion Hamburg)* goes beyond the existing metropolitan region of Hamburg and has a large variety of topics, e.g. networking between institutions for higher education and increased qualification in engineering as well as in natural sciences by co-operations between educational institutions and regional business units. Two other projects focus on the improvement of infrastructure – transport infrastructure to Denmark and the chances associated with this project, as well as better connections between rural areas and cities within the region. The strengthening of clusters in the maritime economy, logistics and life sciences is

1 This is done together with Raum & Energie in Wedel near Hamburg providing the project management.

a further topic. Regional chains of value added and the marketing of regional products are also on the agenda. Beyond that, marketing activities and the use of the touristic potentials, in this case monuments and parks are planned for the next two years. Important is also a development concept dealing with the question on the perspective of the co-operation beyond 2009, including the search for a governance structure and for future projects.

The co-operation between Berlin, Brandenburg and eastern Mecklenburg-Western Pomerania (*Kooperation und Vernetzung im Nordosten*) focuses on the networking of several branches, namely energy, engineering and metal processing, food industry and logistics. An analysis tries to detect strengths and weaknesses as well as requirements of the existing business units. Also, the positioning of the region within Europe and especially within the Baltic Sea-Adriatic Sea-corridor, as well as the better matching of transport infrastructure within the region, are intended goals. The region wants to strengthen its position at the intersection between northern, eastern and central Europe. One project comprises the generation of new topics for the further co-operation in order to become a region of mutual responsibility between urban and rural, central and peripheral, as well as strong and weak regions as demanded by the guiding principle.

The Sachsendreieck (*Metropolregion Sachsendreieck*) with the core cities of Leipzig, Dresden and Chemnitz wants to improve its existing co-operation, mainly an urban network in Saxony. Now, also cities in Thuringia and Saxony-Anhalt, as well as rural and urbanized areas, will be better integrated, and this indicates the need for an appropriate governance model. Furthermore, the networking of research institutions and the economy will be improved by an internet platform facilitating the distribution of information, e.g. for potential partners.

Frankfurt and Mannheim (*Wissen • Kooperation • Innovation Frankfurt/Rhein-Main – Mittelhessen – Rhein-Neckar – Westpfalz*) want to position their two metropolitan regions, including the more rural and urbanized surrounding areas, within knowledge-based society. Analyses on core competences, strengths and weaknesses, and existing linkages in life sciences are the starting points for improving the networks and finding new partners within the region. The regional spatial planning agencies organize platforms and conventions for a further exchange.

The metropolitan region of Nuremberg (*Europäische Metropolregion Nürnberg*) focuses strongly on their rural regions and the potentials there. The formation of clusters in the field of energy is a major goal, as this is a core competency in the regions around Nuremberg, and the involved actors intend to find new partners and new markets. Furthermore, regional chains of added value and the marketing of locally produced foodstuff, as well as local handicraft and local financial service providers, is a project in this MORO. Additionally, the cross border co-operation with regions in the Czech Republic in the fields of economy, research, and administration is a featured topic.

The metropolitan region of Stuttgart (*Metropolregion Stuttgart*) and the surrounding planning associations work on a joint regional development concept,

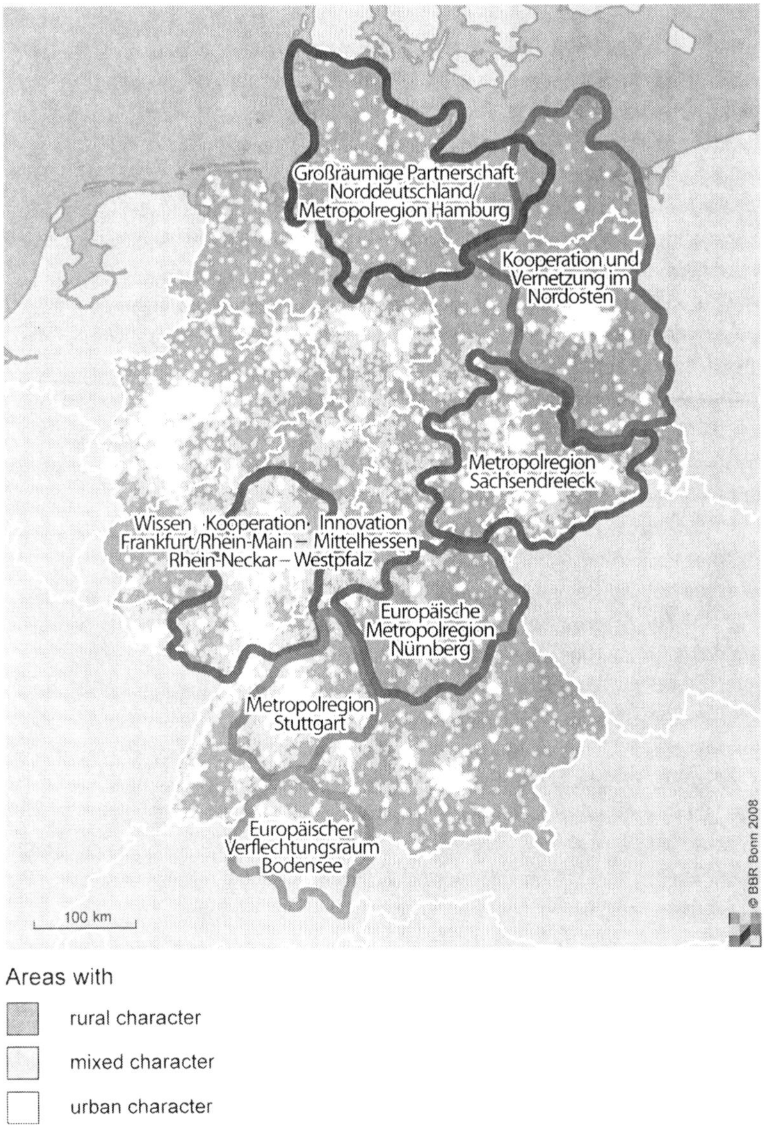

Fig. 4: Model regions of the demonstration project "Supraregional partnership".

as the functional region goes beyond the geographical limits of the individual regional planning associations. But the actors deal not only with win-win-situations, they also want to discover conflicts and try to solve them. Another topic is the better networking in the fields of virtual engineering and of design, as this is a major strength in the region. Regional venture capital to promote new businesses is a third demonstration project.

The Lake of Constance area (*Europäischer Verflechtungsraum Bodensee*) is a cross border co-operation between Germany, Switzerland, Austria, and Liechtenstein. Based on an improved spatial monitoring with statistical data, the actors want to find an appropriate governance structure for better decisions concerning spatial matters and spatial planning.

The seven demonstration regions differ in their spatial structure: Except for the Lake of Constance area, all demonstration projects cover metropolitan regions. All of these metropolitan regions extend beyond their existing spatial limits, e.g. Frankfurt has a joint project with the metropolitan region of Rhine-Neckar, Hamburg comprises also the northern part of Schleswig-Holstein and the western part of Mecklenburg-Western Pomerania – both of them are not included in the metropolitan region of Hamburg. Berlin and Brandenburg, forming a metropolitan region, include the eastern part of Mecklenburg-Western Pomerania. The European metropolitan region of Stuttgart is conducting a project together with the surrounding planning associations, and the metropolitan region of Nuremberg slightly extends its already spacious demarcation, as the Euregio Egrensis also takes part in the project. But despite some differences in the spatial structure – some are monocentric or polycentric, some comprise none, one or two metropolitan regions – the demonstration regions share one element: all include both urban and rural areas as well as central and peripheral ones with different degrees of economic performance. Altogether, the seven model regions cover about 52 % of the German territory, and roughly 48 % of the population in Germany live there.

Supraregional Partnerships in the Light of the Above Observations

In chapter 5.2, some observations have been made on urban-rural partnerships in general. Although the demonstration project "supraregional partnership" is just at the beginning of its work, the question needs to be raised how the projects appear in the light of these observations.

The demonstration project "supraregional partnership" is an implementation of the German guiding principles for spatial development. They go, of course, hand in hand with the Territorial Agenda of the EU, and thus, they are also an implementation of the Territorial Agenda. First results from this demonstration project have already contributed to the European discussion, e.g. in the EU-working group "Coordinating, monitoring and assessing the implementation of the action programme for the implementation of the Territorial Agenda of the EU", Action 1.1 "Prepare and promote policy options between spatial and urban development in the light of the Territorial Agenda and the Leipzig Charter at EU and MS level". Thus, the demonstration project regards itself also as a contribution to the discussion on the European level.

Even though the regions see the difficulties in creating urban-rural partnerships, the broad variety of projects – all in all 30 – and the many motivated actors involved show that this is an appropriate approach for spatial development. The pilot projects want to encourage the feeling for a mutual responsibility between different types of regions and to promote the knowledge that the regions belong together functionally.

This can be best done with projects aiming at win-win situations. But there is the need to find topics for co-operation, also after the funding period, including conflict management, as well.

Another indication for the motivation of the actors can be shown in figures: The seven demonstration regions altogether receive 495,000 Euro for a period of two years. The federal government subsidizes them with about 50,000 to 80,000 Euro for each region. Although this is not a high sum, about 60 initiatives competed for this money. This can be interpreted as a great interest in the topic. Furthermore, the regions themselves add considerable financial means to the project: Altogether, they spend about 619,000 Euro for this MORO from their own pockets, and this is also proof that they regard the money as a valuable investment and the topic as very promising.

There is one additional interesting aspect concerning the most suitable initiatives: Most of the selected examples for the short list of initiatives were initiated by larger cities, less by rural places. This shows that the cities have well understood the mutual relation between the urban and the rural areas. The cities regard the rural and also peripheral hinterland as parts of their functional regions and believe that all regions can contribute with their potentials and functions to economic growth within the larger spatial entity. Furthermore, the importance of rural areas can be seen from the fact, too, that many projects exist which deal especially with rural potential, e.g. the marketing of regional products.

As said above, urban-rural relationship is not yet urban-rural partnership. This is also true for the demonstration regions, although they regard the search for appropriate governance models as an important task. The seven demonstration regions are at different stages in this process: e.g. the metropolitan region of Nuremberg already has a form of organization that places the cities and the rural areas at the same eye level in the political boards of the metropolitan region. Also, the International Lake of Constance Conference has long before existed as a cross border institutional body. But other demonstration regions go far beyond their metropolitan regions, like e.g. Hamburg and Berlin, Brandenburg and Mecklenburg-Vorpommern. Another way to organize the co-operations is by means of finding integrated regional development plans, as done in the region of Stuttgart. These examples show that there is more than one good solution on how to incorporate rural areas into the projects and how to form the basis for further decisions. Thus, the demonstration project is also an experiment for discovering those factors promoting and constraining the emergence of appropriate governance structures.

5.4 Conclusions

Urban-rural partnerships are an important element for spatial development. They provide a way to combine growth and balancing policy and they enable peripheral and economically weak areas to participate in growth and innovation. But this will only work if all parts contribute with their potentials to economic growth. The idea is evident; nevertheless, there are still open questions on how to achieve these goals. Who are the relevant actors? How to encourage the relevant regional actors and especially the economy to participate in this process? What are topics or projects of shared interest for these partnerships? What are adequate governance models? The observations from Europe in general and from Germany in detail have provided an overview of the differences between existing projects, their shortcomings, but also of the importance of promoting the idea of urban-rural partnerships. The demonstration project tries to answer these questions also with the help of the experience in the demonstration regions, showing that urban-rural partnerships can work and are regarded as a solution to combine growth and balancing policy in larger regions.

References

Baltic+ (2004): Concept for development of spatial partnerships between contrasting areas. n.f.i. http://www.raumpartnerschaften.tu-berlin.de/datpdf/eKonzept.pdf (retrieved on 15.09.2007)

Bengs C., Zonneveld W. (2002): The European Discourse on Urban-Rural Relationships: A New Policy and research Agenda. In: Built Environment Vol. 28, No. 4, 278-289.

Bundesamt für Bauwesen und Raumordnung (2001): Study Programme on European Spatial Planning. BBR-Forschungen No. 103.2. Bonn.

Bundesamt für Bauwesen und Raumordnung (2005): Raumordnungsbericht 2005. Report, No. 21. Bonn.

Comhairle Nan Eilean Siar (2002): Western Isles Structure Plan. lLand Use Strategy for Sustaining the Communities of the Western Isles. Stornoway. http://www.cne-siar.gov.uk/develop/pilot/structureplan.pdf (retrieved on 24.09.2007)

Council of European Municipalities and Regions (2003): Complementarities between urban and rural areas in promoting employment and social inclusion. Brussels. http://www.ccre.org/bases/T_599_8_3524.pdf (retrieved on 09.09.2007)

DG Regional Policy (2004): Interim Territorial Cohesion Report. Luxembourg. http://ec.europa.eu/regional_policy/sources/docoffic/official/reports/coheter/coheter_en.pdf (retrieved on 28.11.2007)

European Commission (1999): ESDP European Spatial Development Perspective. Towards Balanced and Sustainable Development of the Territory of the European Union. Luxembourg. http://ec.europa.eu/regional_policy/sources/docoffic/official/reports/pdf/sum_en.pdf (retrieved on 28.11.07).

Expertenkommission „Überprüfung und Neukonzeption der Regionalpolitik" (2003): Neue Regionalpolitik (NRP). Final report. http://www.news-service.admin.ch/NSBSubscriber/message/attachments/8095.pdf (retrieved on 15.09.2007)

Gatzweiler H-P (2006): Leitbilder in der Praxis. Impulse zu Umsetzung durch Modellvorhaben der Raumordnung. In: Informationen zur Raumentwicklung No. 11/12, 677-691.

Government of Romania (2005): Government Program of Romania 2005-2008. http://www.guv.ro/engleza/obiective/afis-docdiverse-pg-eng.php?iddoc=21 (retrieved on 02.10.2007)

Informal Ministerial Meeting on Urban Development and Territorial Cohesion (2007): Territorial Agenda of the European Union. Towards a More Competitive and Sustainable Europe of Diverse Regions. http://www.bmvbs.de/Anlage/original_1005295/Territorial-Agenda-of-the-European-Union-Agreed-on-25-May-2007-accessible.pdf (retrieved on 28.11.2007)

Kawka R. (2007a): Wachstumsregionen in Deutschland – empirische Befunde. In: Köhler S. (ed.): Wachstumsregionen fernab der Metropolen. ARL-Arbeitsmaterial No. 334, 36-50. Hanover.

Kawka R (2007b): Typisierung ländlicher Räume – Stärken, Schwächen und Funktionen. In: B&B Agrar, No. 3, 102-105.
Metropolitan Areas Baltic Sea Region (2006): Polycentricity in Practice. n.f.i.
Nova-Institut (2006): Regionen Aktiv Newsletter XVI, October 2006.
OECD (2001): OECD Territorial Reviews Hungary. Paris. http://wbln0018.worldbank.org/Apps/CCKDoclib.nsf/fd3f251023ca1bdc85256896006bf440/8659b41e6d944a5b8525708100710ef30/$FILE/Territorial%20Reviews%20(Hungary).pdf (retrieved on 24.09.2007)
Polverani L., Quiogue N., Gross F., Novotny V. (2005): Territorial Cohesion and Structural Funds Programmes: Urban Development and Territorial Cooperation. European Policies Research Centre IQ-Net Thematic Paper No. 16 (2). Newcastle upon Tyne. http://www.eprc.strath.ac.uk/iqnet/downloads/IQ-Net_Reports(Public)/16.2Urban.pdf (retrieved on 05.11.2007)
Scott Wilson Business Consultancy (2006): Sustainability Appraisal of Newham's Local Development Framework. Sustainability Report Technical Appendix: Context and Baseline Information. London. http://www.newham.gov.uk/nr/rdonlyres/820dedd8-660a-4878-b22c-20b7f1b5bfe8/0/sus_appr_tech_con_base.pdf (retrieved on 24.09.2007)
Senatsverwaltung für Stadtentwicklung, Ministerium für Landwirtschaft, Umweltschutz und Raumordnung, Gemeinsame Landesplanung Berlin-Brandenburg (2007): Gemeinsames Landesentwicklungsprogramm der Länder Berlin und Brandenburg (LEPro). Berlin.
Standing Conference of Ministers Responsible for Spatial Planing (2006): Concepts and Strategies for Spatial Development in Germany. Berlin
Steineke J. (2007): The impact of recreational homes. In: Journal of Nordregio, Vol. 3, No. 7, 10-11.
Zimmermann H. (2007): Anmerkungen zum Argument der Verantwortungsgemeinschaften. In: Raumforschung und Raumordnung, No. 2, 213-216.
http://www.regionalmarke-eifel.de (retrieved on 11.06.2008)
http://www.unser-land.info (retrieved on 11.06.2008)http://www.vgrdl.de (retrieved on 20.05.2007) Bruttoinlandsprodukt, Bruttowertschöpfung in den kreisfreien Städten und Landkreisen Deutschlands 1992, 1994 bis 2006.

6 The Importance of Leitbilder for Structural Change in Small Towns

Gerd Lintz, Peter Wirth

Introduction

The "Visions and Strategies for Spatial Development in Germany"[1] have provoked new discussion on Leitbilder in spatial development.[2] They are seen in the tradition of German spatial planning, the purpose of which is to create desirable spatial structures. New Leitbilder are a response to globalisation, to the transformation processes induced by the unification of Germany, to demographic change, and to the eastward enlargement of the European Union.[3]

In general, Leitbilder have begun to play a more important role in German spatial planning, at the urban level[4], in regional development[5], and in federal spatial planning (see the Spatial Planning Policy Guidelines) since the 1990s. Regardless of the spatial level at which they are situated, Leitbilder are special superordinate objectives which have many functions to perform.[6] Internally, they serve to guide and coordinate action and to activate and motivate the parties involved. Externally, Leitbilder can shape the image of a city, town, or region, and thus have a marketing function. They are a common element of strategic planning.[7]

In major cities, and especially in international metropolises, such overall concepts are indispensable. They raise the profile of a city.[8] Cities face international or even global competition, and ultimately represent not only themselves but entire urban regions. If they are to hold their own against competitors, they must be "well equipped." They therefore need Leitbilder that equally promote internal cohesion, focus activities, and market the "location". But what is the situation with small towns?[9]

This category of urban centre, the problems of which are typically overshadowed by the debate on cities and metropolitan regions, is also under considerable pressure

1 BMVBS (2006)
2 cf. Lutter (2006) for a clear indication of this
3 Lutter (2006)
4 Becker, Jessen and Sander (1998)
5 Hein (1998)
6 Dehne (2005)
7 Dehne (2005), Knieling (2006)
8 cf. Jessen (2005)
9 This is defined for our purposes as settlements with a population of between 5,000 and 20,000.

for change. The factors involved include growth and shrinkage, aging, social segregation, the quality of life in the inner city, land consumption, brownfields, and traffic. The first thesis states that Leitbilder could facilitate orientation, especially in critical processes of structural change. In decision-making processes that are complex and difficult to asses, they could help to provide focus on new perspectives and to mobilise actors (internal function). And they could spotlight unique characteristics and thus improve the image of the town (external function). In this sense, Leitbilder are to be understood as elements of an active urban development policy.

Although few authors in the academic discussion reject Leitbilder as strategic planning tools in general terms, there are critical points and "dangers" that ought to be taken seriously, not least of all when considering small towns with little governing capacity. Leitbilder are accused of lacking content, of being authoritarian and inflexible.[10] Sieverts points out that if such overall concepts are treated as dogmas in urban development they endanger flexibility and creativity.[11] Dörner even claims that rigid adherence to Leitbilder can provoke "catastrophes".[12] A counter-thesis to a Leitbild-driven planning style in uncertain situations is to "muddle through" without any such guiding concept.[13] Especially in periods of radical change in which orientation is absent, a situational, reactive, step-by-step procedure is conceivable or perhaps even advisable. It could also provide motivation in the absence of medium and long-term visions.

Against this background, we address the following questions. How do Leitbilder come about in small towns? What forms do they assume? What functions do they perform? The answers will help in dealing with the superordinate question: What role do Leitbilder play for structural change in small towns? The method adopted is that of comparative case-study analysis of three small mining towns in eastern Germany which were forced to cope with extreme structural change following the unification of Germany in 1990. Such communities offer particularly useful insights into the importance of a Leitbild approach for the development of small towns.

Section 6.2 addresses certain basic considerations, explaining the Leitbild concept and the structure and function of small towns in greater detail. Section 6.3 discusses the empirical findings on the three case examples. Section 6.4 provides a comparative examination of the various roles played by Leitbilder in the settlements under study. Finally, section 6.5 presents conclusions on the importance of Leitbilder for structural change in small towns.

10 cf. Dehne (2005)
11 Sieverts (1999)
12 Dörner (2006)
13 Lindblom (1959)

6.1 Basic Considerations on Leitbilder and Small Towns

In view of the broad discussion on Leitbilder and the complex reality behind the term "small town," we begin with certain basic considerations concerning the conceptual and substantive framework for describing and discussing the case examples.

The Role of Leitbilder for Spatial Development

With reference to Dehne, a Leitbild for spatial development is to be understood as "a clear, superordinate objective for an area … that can be accepted by the majority of the people and institutions affected, guide the activities of individuals relating to the area in question, and which can hence steer spatial development".[14] Leitbilder are to be found at all levels of spatial development (local, regional, national, European, global). They must satisfy certain requirements if they are to be effective. They are comprehensive in nature, steer spatial processes in society, set standards, and are abstract, ideal-typical, clear, and consensual.[15] Generalised Leitbilder for spatial development (e.g., sustainable regional development, decentralised concentration, the compact city) are to be distinguished from individual, area-specific Leitbilder (e.g., "Metropolis Hamburg – Growing City," "Capital Region Berlin-Brandenburg"). This paper focuses on the latter.

Leitbilder can assume various forms. Ideally, they are to be found in textual and pictorial form, as in the Visions and Strategies for Spatial Development in Germany.[16] Their significance can be stressed by issuing them in attractive brochure form. But Leitbilder are also to be found as components of programmes, concepts, and plans, not least of all in regional development strategies, which have enjoyed a boom in Germany and Austria since the 1990s.[17] Finally, Leitbilder can exist as powerful ideas without being put down on paper. This form of Leitbild has largely been ignored by literature. It is propagated by word of mouth by proponents in influential positions who think in strategic terms and pursue certain goals. The chapters to follow examine how the various types of Leitbild gain persuasive power.

In general, spatial Leitbilder are expected to perform the following functions. First, they offer medium to long-term orientation for the actors involved. This is particularly important in times of radical change and uncertain perspectives. Second, they have a coordinating function. This is necessary where – typical for urban and regional development – interested parties from a wide range of fields and levels collaborate in pursuing common goals. Third, Leitbilder have a motivating function. They aim to stimulate development and innovation processes in which many actors can play a part. Fourth and finally, they are designed to polish the image of cities and regions and to position them in competition with others, an

14 Dehne (2005: 608)

15 ibid.: 609f.

16 BMVBS (2006)

17 Diller (2004)

important task as competition between locations grows.[18]

These introductory remarks suggest that Leitbilder can be pivotal in determining the future options of spatial entities. The main objective of this article is to show the extent to which this complies with the situation of small towns. Before dealing with specific examples and going into detail, certain characteristics of small towns need to be considered which determine the local setting for framing and operating with such models.

Characteristics of Small Towns

In the context of urban development, the planning sciences argue about whether small and medium-sized towns are "miniaturised blueprints of the city" or whether they are largely different in nature.[19] Although the empirical basis for such conjectures is thin[20], certain aspects tend to support the thesis that such settlements are different in character. To some extent they also have a bearing on medium-sized towns; we however focus our interpretation according to their relevance for the "small town."

Small towns have limited general functions. They are sparingly endowed with infrastructure. Particularly lacking are high-quality facilities like theatres, specialised hospitals, and leisure centres. But, depending on location and size, they perform important functions for their catchment areas, and much of the demand for services comes from the surrounding region. Particularly in rural areas, they perform supply functions and ensure the stabilisation of settlement structures. Small towns are thus an important element in the polycentric urban system.[21] Since the attractiveness of small and medium-sized towns is increasingly determined by quality-of-life factors like residential environment, infrastructure, public security, and educational opportunities,[22] careful attention must be paid to such factors.

The importance of the small town as a location for business and industry is in decline. In recent years, there have been complaints in Europe about the loss of economic functions for small towns.[23] Since jobs are being lost, working interrelations in the interactional space are declining (commuting). This calls for adequate reaction and, as a rule, for new orientation, either to maintain the status quo or to develop options for the future. Not only a town's material resources are important, but also the capabilities of actors and the prevailing style of control.[24] And not only local politics and administration are involved, but also business.[25]

The capacity of local authorities is limited. Unlike cities with a well-developed sectoral administration, small towns have little administrative capacity for planning

18 cf. Dehne (2005: 610 f.), Koch (2006: 58 ff.)
19 Baumgart et al. (2004)
20 including ÖIR (2006)
21 Adam (2005)
22 Erickcek and McKinney (2006)
23 Courtney and Errington (1999)
24 Bataïni et al. (2002)
25 Andersson and Karlsson (2004)

and control. The smaller the town, the more administrative staff has to play multiple roles. Particularly regarding questions of urban development, the smaller a city, the less capability it possesses in order to tackle complex tasks on its own. Many small towns have to give up on major projects because they do not have the necessary administrative resources or they have to rely on external assistance.[26]

In contrast, their size guarantees that they can be well managed and comprehended. The greater comprehensibility of institutional and non-institutional structures in small towns is considered an advantage. It permits more detailed knowledge of problems, greater flexibility in taking action, shorter lines of communication, and more direct contact between actors.[27]

The local milieus play an important role. Spatial proximity and personal contacts enable the development of stable local milieus, permitting a rapid response to changes in underlying conditions and innovative strategies.[28] Many small towns thus have good human resources at their disposition and have developed modes of decision-making that can be superior to those in large urban centres, which can be more sluggish in regard to purposefulness. However, such networks can also prove to be "preventive alliances," blocking innovation.[29]

As these characteristics show, the development resources available to small towns can both pose problems and offer opportunities. Owing to the diversity and functional specialisation of small towns, no uniform profile regarding strengths and weaknesses can be established.[30] This suggests that Leitbilder for spatial development of small towns will be very specific and individual.

6.2 Empirical Findings

Investigative Framework

The empirical studies examine small towns in eastern Germany that have experienced far-reaching structural change since 1990. All are former mining towns, the economic basis of which, mining and associated industries, collapsed in a brief time span. They faced a critical situation that can be summed up by using keywords such as economic monostructure, extensive environmental damage, high unemployment, and outmigration. Under these conditions, considerable pressure for change obviously built up, calling for adequate concepts and strategies. In addition to the challenges faced by the whole of eastern Germany in the early 1990s, the small towns under study had to, more or less, entirely rebuild living conditions and an economic basis. A new identity had to be created.[31] These extreme cases provide

26 Schäuble (2004)
27 Rüdiger (2004)
28 Baumgart et al. (2004)
29 Grabher (1993)
30 Aehnelt et al. (2006)
31 Leimbrock (2000)

particularly useful insights into the role of Leitbilder.

The approach adopted was a comparative / multiple-case study design[32]. The underlying studies were conducted in the course of the READY (Rehabilitation and Development in Mining Regions) project under the EU Community initiative INTERREG III B between 2003 and 2007. The participants in this complex project, involving partners from six countries, included seven small towns in eastern Germany which provide the basis for our investigations: Bad Schlema,[33] Gräfenhainichen, Hartenstein, Johanngeorgenstadt, Lugau, Oelsnitz/Erzgebirge, and Schöneck/Vogtland (Figure 1).

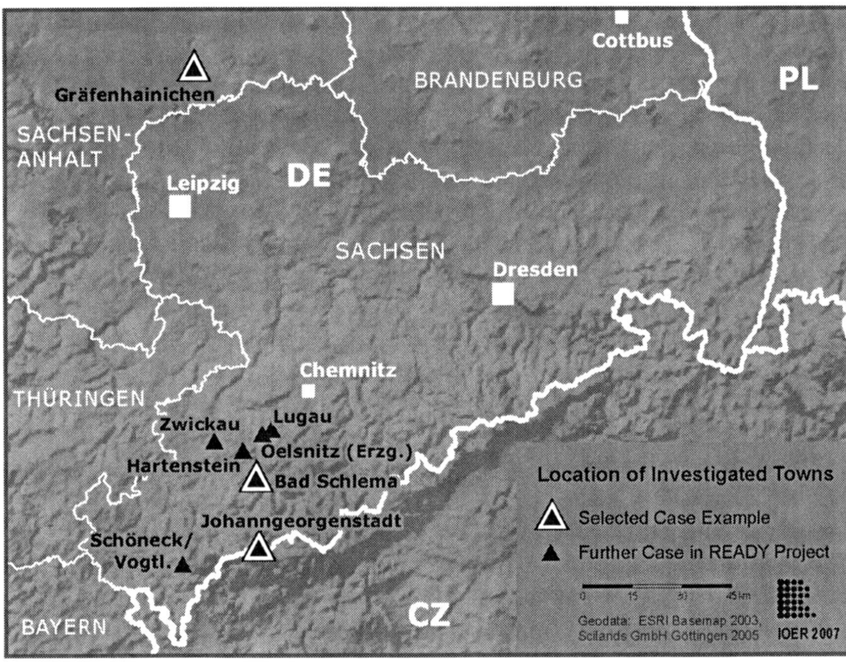

Fig. 1: Case studies in eastern Germany.

Data was collected in each town in several steps. In the first step (initial survey), statistics and local government documents were examined and structured interviews were conducted with key actors in order to prepare a "profile." The second step involved sporadic participant observation, e.g. at workshops and meetings of bodies

32 Yin (2002)

33 Bad Schlema was the only community among those under study without town status (Stadtrecht). But by size, the predominance of urban structures, and the industrial past justified treating Bad Schlema on par with small towns.

by the participating investigators over a period of three years.[34] In order to close remaining gaps in knowledge and obtain background information, surveys were repeated in the towns. The period under study was 1990 to 2006, from the collapse of the GDR, which triggered a far-reaching process of change in the researched towns, to the present day.

Empirical studies in the READY project sought primarily to determine the factors for successful structural change. They also examined whether the towns had Leitbilder at their disposition, and what role they played in the local development process in the eyes of local actors. Only in the course of these investigations did it become clear that the results would also yield interesting findings on the use of Leitbilder. The relevant information was processed separately for the present study. It proved that three of the seven cases under study best reflected the possible spectrum of Leitbild development and implementation in the small towns affected by structural change. These three cases are therefore most appropriate for answering the research questions. At the same time, the authors were well acquainted with these examples from other research projects, so that they had far-reaching knowledge about problems and approaches in the towns and the progress achieved in development.[35] Additional telephone interviews were conducted to update knowledge.

The cases researched in greater depth include the former mining towns of Bad Schlema (Saxony), Gräfenhainichen (Saxony-Anhalt), and Johanngeorgenstadt (Saxony).

Bad Schlema – From "Valley of Death" to Health Spa Resort

In the Erzgebirge community (pop. 5,600) near Chemnitz in Saxony, 44 years of uranium mining by the Soviet-German Wismuth Corporation were brought to an end in 1990. Practically the entire community was in a disastrous state. Appalling mining damage and radiological hazardous waste induced the tabloids to describe Schlema as a "Valley of Death." In the heart of the suburb Oberschlema there was a 25 ha "deformation area," which had been off limits since 1970 owing to the hazards posed by underground cavities. Although some miners were employed in remediation, many jobs were lost with the cessation of mining activities in Schlema.

Only 15 years later, Bad Schlema presents itself as a state-approved radon health spa resort with the slogan "Naturally Schlema." Guests are offered a broad range of leisure activities in one of the "most attractive and modern health resorts in Saxony," including a daily programme at the "Actinon" health pool, in the resort hotel, in the "Aktivist" mineworkers' cultural centre, in the exhibition mine, and elsewhere. Guests can bathe in the warm radon and salt water pools, enjoy a sauna, generally relax, or take a cure, far away from the stress and hectic pace of everyday life. The summary of local history on the town's website describes Bad Schlema's

34 In this phase the authors were joined by Katharina Mörl and Leander Küttner from the IÖR in collecting data.

35 cf. Leimbrock and Lintz (2002), Wirth (2003)

transformation from a uranium mining community to a radon spa between 1990 and 2006.

Immediately after mining ceased in 1990, the idea was born to reconstruct the radon health spa, thus giving the community a new identity. The spa had made Oberschlema famous throughout Germany in the first half of the 20th century, only to be completely destroyed by mining activities later. The mayor and a small group of close associates, although accused of pipe dreaming, had enough authority and persuasive power to convince the population. The idea became an as yet unwritten Leitbild. A scientific study showed that the radon waters in Schlema still had healing properties. The project was then tackled with far-reaching participation by local actors (some 190 private persons are shareholders of the resort company) and strong financial support from the European Union, the federal government, and the Saxony state government, and was largely in place by 2006. Environmental damage has for the most part been remedied. The town is embedded in a harmonious "remediated landscape."

The idea of reconstructing the health spa, which had come up immediately after the change of regime, had been integrated as early as 1992 into the rehabilitation plan of the government-owned Wismut GmbH, in charge of mining remediation. In 1995, it was presented in the health resort development concept in the form of a brief written Leitbild. Schlema was described as a "green midpoint between Aue and Schneeberg." The paramount objective was to obtain official designation as a "*Heilbad*", a health spa resort. In the years that followed, almost all local authority efforts were engaged in implementing the concept. In 1998, the spa house and the spa park over the former deformation area were opened. Additional impetus was gained from participation in the Hanover EXPO 2000 as a decentralised project. In 2004 Schlema was designated a "radon health spa resort." At present the overall project is being rounded off with additional elements (including a golf course).

What is remarkable about the case of Bad Schlema is how fast it set a vision for the future after the change of political regime, and how consistently it was implemented. The overall concept of "reconstructing the health spa resort" proved highly convincing, owing primarily to the extraordinarily strong personality of the mayor, who comes from a mining background. The fact that this Leitbild was given written form played a subordinate role. All important projects were directed towards implementing the overall concept anyway. It determined the internal orientation and coordination of the development process.

Gräfenhainichen – "Town with New Energy"

Gräfenhainichen is located in Saxony-Anhalt, some 50 km north of Leipzig in Wittenberg County. The town is served by two federal highways and a railway. 1990, the year of reunification, marked a breach in the development of the town. Extensive opencast lignite mining came to an abrupt end, leaving two vast surface mining areas to the north and south, covering some 2,500 ha, in need of remediation. The central workshop for opencast mining equipment was closed down. The population declined from 10,382 in 1987 to 7,990 in 2004. The rate of unemployment in 2003 registered at the Gräfenhainichen agency for employment was some 22%.

Today, the town presents itself on its website with great self-confidence as a "small town of great variety," well on its way to becoming a "charming small town." The factors emphasised include the printing tradition (there is a small museum) and the town's proximity to the Dübener Heath recreation area. It is recalled that Gräfenhainichen is the birthplace of the famous hymn writer Paul Gerhardt, that it is the location of the "excavator city" Ferropolis, and that the town has endowed itself with the title "Town with New Energy." In 1990, no one would have foreseen this new town image.

After the shock caused by industrial decline from 1990 to 1992, the unwritten Leitbild of Gräfenhainichen comprised two obvious goals: attracting new industry to compensate for the job losses in mining and rehabilitating the historic town centre, which already featured a pedestrian zone during the GDR era. Rehabilitation of the town core was strongly supported by the Saxony-Anhalt state government and can now be considered completed.

In contrast to operations in the southern sector, the cessation of opencast mining to the north had been decided even before 1990. The final working plan provided for flooding the opencast workings; Gräfenhainichen was accordingly to become a lakeside town. From 1992, the systematic mining remediation by the Lausitzer und Mitteldeutsche Bergbau-Verwaltungsgesellschaft was influenced by the Bauhaus Dessau, the creative school of applied arts that had contributed much to classic Modernism in the early 20th century. The Bauhaus took an intense interest in designing the post-mining landscape in central Germany, and introduced the idea of setting up a "city of iron" (Ferropolis) on a peninsula in the envisioned northern lake, in which an ensemble of five giant opencast mining machines would form the nucleus for further development. As the prospects for attracting industry waned, hopes were increasingly set on tourism, culture, recreation, and quality of life as factors in developing the future of the town – with Ferropolis as a highly visible advertisement for the project. Given the assets favourable for tourism and the proximity to the Bauhaus in Dessau and to Wittenberg, the city of Martin Luther, this appeared to be a realistic option. Many activities and projects were directed towards implementing this vision.

The Gräfenhainichen municipality took a courageous step in 1994 in acquiring the site from the Treuhand Agency. In 1995, the excavators were installed and the city symbolically founded. From 1995 to 2000, plans were pursued as a decentralised project of the Hanover EXPO 2000 (correspondence region Saxony-Anhalt). The municipality set up the Ferropolis GmbH as a development agency, and a sponsoring association was founded. A consistent overall concept for the town took shape, and the North-South axis, which traversed the historic centre, became a link to be developed between the two future lakes. At the northern end of this axis, a so-called "town balcony" was erected in 2001, providing a view across the lake to Ferropolis and opening up the town to the lakeside.

While the idea of Gräfenhainichen as a residential town and tourist centre with small and medium-sized businesses in an attractive landscape was further developed and strongly integrated regionally (e.g., Ferropolis Park), a comprehensive, written Leitbild was systematically elaborated with public participation in the context of the Stadtumbau Ost ("Urban Redevelopment East") competition. A further important aspect was added: energy efficiency and the far-reaching use of renewable energy. The motto "Town with New Energy" alludes both to a past determined by lignite (old energy) and the will of the town to make every effort to cope with structural change. First projects for using near-surface geo-thermal energy have already been implemented.

Overall, the present Leitbild for Gräfenhainichen has slowly developed and taken shape from a range of elements since 1990. A high degree of effectiveness has been evident at all stages. For a small town, the Leitbild shows extraordinarily high conceptual quality. This could be achieved only through great openness on the part of the municipality and intensive cooperation with the Bauhaus Dessau and a former Bauhaus contributor.

Johanngeorgenstadt – Centre of a "Sports, Recreation, and Health Region"

The old mining town of Johanngeorgenstadt in the Erzgebirge, like nearby Bad Schlema, was radically transformed by uranium ore mining. In contrast to Schlema, mining operations in Johanngeorgenstadt lasted only 13 years (1946 to 1959) before the deposits were exhausted. In spite of this relatively brief period of ore extraction, most of the town had to be demolished owing to undermining. The original population of 10,000 grew to 40,000 in the 1950s, only to decline to its former level little later. Now only some 5,000 people still live in Johanngeorgenstadt. What remains is a community comprising many mining settlements scattered across the landscape. Although over 400 suspected contaminated sites were still listed after 1990, the town did not fall under the so-called Wismut Act, and until 2003 received no regular government assistance with remediation. For this and other reasons, (peripheral location, difficult access, decline of manufacturing) the population declined by almost half since 1990.

On its website, Johanngeorgenstadt presents itself as a state-approved recreation resort and "Town of the Candle Arch," a Christmas decoration of local origin. With the Nordic combination Olympic medallist Björn Kircheisen and the nationally renowned traditional music group "De Randfichten," the town has two prominent advertising vehicles. Johanngeorgenstadt displays a tourism profile in the media that stresses sporting and exercise activities (e.g., skiing, hiking, climbing, skating). The town is surrounded by large forested areas with bizarre rock formations and raised bogs.

Since 1990, Johanngeorgenstadt has experienced much local political turbulence, bringing a succession of mayors and shifting majorities in the town council. In the early 1990s the unwritten Leitbild of reconstructing the historic centre that had fallen victim to mining operations emerged. This first guiding idea after the change in regime has still not been realised owing to the lack of investment and political support. To get planning under way, a town planning concept was drawn up between 1995 and 1997. The point of departure was careful downsizing of the barrack-like uranium mining estate. The remnants of the historic core and the new town built in the mid-1950s were to be developed into sub-centres. Although the basic problem of further development was recognised and a planning concept had been elaborated, there was one fundamental flaw. The necessary remediation of uranium mining damage was not taken into account. This made implementation of the concept quite impossible.

As a consequence, the town gratefully took up the offer of the state government of Saxony in 1997 to provide Johanngeorgenstadt and six neighbouring local authorities with substantive organisational and conceptual assistance in the framework of a cooperation project titled "Uranium Mining Remediation and Development Area." With external coordination, working groups for remediation, tourism, commerce and industry, and accessibility were set up step by step, which involved not only the local authorities but also the county, interest groups, and associations.

In 2000 these joint efforts resulted in a comprehensive written Leitbild under the heading "Sports, Recreation, and Health Region", designed to overcome the consequences of uranium mining, to improve the quality of life, and to ensure balanced economic and ecological development. Action was envisaged in remediating mining damage, improving accessibility, creating new jobs, developing tourism, town redevelopment, and maintaining the infrastructure. The agreed principles for action were participation, intermunicipal cooperation, and cross-border cooperation with neighbouring Czech local authorities. The Leitbild is part of a joint plan of action for Johanngeorgenstadt and six adjacent local authorities.

The greatest cooperative success achieved was to obtain funding for remediation from the federal and the state government of Saxony in 2003 for the period up to 2012. This eliminated a major obstacle to regional development. But joint implementation of the development model nevertheless failed because no common position could be agreed on between the involved local authorities regarding basic development issues, such as safeguarding schools and cooperation between local housing companies. Nonetheless, Johanngeorgenstadt still pursues the tourism components of the Leitbild. The town cooperates with partners in Germany and the Czech Republic on new tourism schemes, and concentrates on small and clearly defined projects that can be handled with the resources available. The ambitious ideas advanced by the town's various Leitbilder have not been put into effect.

Johanngeorgenstadt therefore stands for the persevering but unsuccessful search for a concept that generates attention. However, the general principles elaborated to date, which had even resulted in an exemplary intermunicipal Leitbild, have failed to exert a decisive effect. The impression is that all major projects have been bogged down in the struggle to obtain government assistance for mining remediation. Instead of motivation, resignation and disappointment became the predominant sentiments among the population.

6.3 Leitbilder in Small Towns as Strategic Planning Tools

Having examined the role of Leitbilder in each of the three small towns selected, we now analyse the emergence, form, and function of such overall concepts in the local development process from a comparative point of view. Leitbilder have been already described as tools for strategic planning (6.1). According to Healey "…'strategic spatial planning' refers to self-conscious collective efforts to re-imagine a city, urban region or wider territory and to translate the result into priorities for area investment, conservation measures, strategic infrastructure investments and principles of land use regulation".[36] It is thus a superordinate concept, concerned not with producing a plan but with attaining agreement on fundamental objectives and implementing them in the context of a social process.[37]

Origins of Leitbilder

The collapse of the GDR in 1990 completely changed the prospects for the three eastern German small towns under study. The setting for economic activity and community development, as well as demographic and political conditions changed overnight. The old development goals were obsolete, and new ones were not yet existing.

It was therefore no surprise that the search for orientation was a major priority in the first years. In Schlema, reconstruction of the health spa resort was envisaged, in Gräfenhainichen the plan was aimed towards re-industrialisation, while Johanngeorgenstadt wanted to rebuild the historic old town. These ambitions constituted a Leitbild, but there was no comprehensive written document. Uncertainty and a sense of urgency prevailed. Initial strategies logically integrated earlier lines of development: Schlema had already been a spa 60 years ago, Gräfenhainichen had developed into an industrial town in the course of a century, and Johanngeorgenstadt looked back on the period before uranium mining when it was a tranquil small town with a winter sport tradition. Ultimately, the initial idea prevailed in Schlema alone.

After this first exploratory phase, lasting 3 to 4 years, the towns either elaborated more specific strategies or began to look farther afield. In Schlema, the prerequisites for reconstructing the health spa resort had been clarified (government remediation programme, support from the federal and state governments, certification of the healing properties of the radon water). In Gräfenhainichen it had become clear that the idea of re-industrialisation alone was not viable. Ferropolis, the city of iron, and the concept of the lakeside town offered viable and promising alternatives which attracted external support (EXPO 2000 correspondence region, Bauhaus Dessau as provider of ideas, government support for projects). In Johanngeorgenstadt, it transpired that neither the investment resources nor state government support could

36 Healey (2004)
37 cf. Albrechts (2004)

be obtained for reconstructing the historic old town. Moreover, no solution had been envisaged for the remediation problem. A town development concept was elaborated in the hope of generating perspectives. In this phase, in which a clear orientation was found in at least the first two cases, comprehensive, written Leitbilder were produced by all three towns. This self-discovery phase lasted between 3 to 6 years.

In the third and final phase of the period under study, strategies were consolidated. Schlema largely completed remediation of mining damage, the health spa became the new and biggest employer and a public attraction in the town, and the community was awarded the title of "health spa resort." Gräfenhainichen continued with urban redevelopment, built the town balcony on the new lake, and, with Ferropolis, gained a new landmark. Only Johanngeorgenstadt was slow in making progress. The first municipal Leitbilder had to be rejected because it was impossible to implement them. A new, regional Leitbild was developed which lacked leverage. A modest strategy of small steps resulted. The findings can be summed up as follows:

- In all cases, Leitbilder played an important role.
- Strategies and Leitbilder proved highly context-dependent, varying in terms of the problems, structures, conjunction of actors, and underlying conditions of the town. The attractiveness of Leitbilder also depended on the potential the towns offered.
- General and vague Leitbilder consolidated increasingly into more concrete and specific, written documents. The combination of several "partial Leitbilder" to form "overall Leitbilder" can also be found.
- Development and implementation of Leitbilder occur simultaneously. They were permanent processes without beginning or end, but with breaks and turning points (due, for example, to new projects and actors).
- The development of Leitbilder and strategies took place in several stages (search phase, self-discovery phase, consolidation phase).

Form of Leitbilder

At the end of the investigation period all three towns had written Leitbilder. And in all cases, they were part of more far-reaching programmes, concepts, or development plans that were exclusively informal in nature. In Bad Schlema it was the health resort development concept, in Gräfenhainichen the town development concept, and in Johanngeorgenstadt a regional plan of action. Explicit Leitbilder like those often published in brochure form for large cities were not prepared in any of the cases under study.

As the account of the generation and development of Leitbilder has shown, they were not yet available in written form at the onset of radical change. However, they were present as unwritten Leitbilder of conceptual character. "We always knew what we wanted. We didn't have to write it down!" as a former mayor of Bad Schlema put it. Similar statements and observations were made with regard to the other examples. Verbal communication about the superordinate goals of urban development apparently sufficed. This is evidenced by the fact that written

Leitbilder were mostly accessible only to a small circle, whereas many more people were involved in strategy development and implementation. These observations permit a number of conclusions.

- Although written Leitbilder existed in all the cases under study, they were of limited importance.
- Typical for Leitbilder is their embedding in planning documents and concepts.
- Verbally disseminated Leitbilder in the minds of key actors are still more important.
- Format or shape do not appear to be decisive characteristics for the effectiveness of Leitbilder.

Functions of Leitbilder

The functions of Leitbilder are the most difficult aspect to demonstrate. For the external observer, they are often only sporadically apparent and therefore mostly have to be identified indirectly. The functions of Leitbilder can be more easily understood if one recognises that the development process in the small towns under detailed study was determined by relatively small groups of key actors, not exceeding between five and seven people. They came from the administration, local politics, business, and the general public. As a rule, they were organised in networks. These actors were backed by organisations and sectoral networks to which they could convey concepts agreed on by the circle of key actors, thus ultimately increasing their reach. The limited dimensions of the small town make it possible to practically reach all creative circles without requiring a particular campaign.

One central function of Leitbilder is to orient key actors towards superordinate objectives. In Schlema, for instance, concentration on development of the health spa played a focal role. In Gräfenhainichen, re-orientation of the Leitbild after the relatively unavailing efforts at re-industrialisation towards the Ferropolis project and tourism had considerable success. In Johanngeorgenstadt, it was tourism above all other goals as constant focus of local initiatives. This function of Leitbilder can accordingly be fully confirmed for the small towns under study. Moreover, the thesis was confirmed that the orientation function was particularly important in the context of the structural paradigm change in the eastern part of Germany.

A second function of Leitbilder is coordination. It is of high importance in attaining the envisioned goals. In Bad Schlema, close cooperation between the health resort company, municipality, and private landlords and service providers played a role. Just as important was coordination with government remediation enterprises, the state government of Saxony, and the federal ministries responsible. Important in Gräfenhainichen was not only close cooperation within the town and with the remediation enterprises, but also collaboration with the Bauhaus Dessau and the Saxony-Anhalt state government. Furthermore, Gräfenhainichen cooperates in a regional scheme entitled "New Lake Country Dübener Heide" designed to promote tourism. Since the mid-1990s, Johanngeorgenstadt has been collaborating intensively with Saxony state planning authorities, and has achieved progress in cooperation with the neighbouring Czech local authority. As the examples show, internal coordination is the easier task, facilitated by the mentioned networks. But at least as important in this regard is external coordination to compensate for lacking capacities: for example coordination and cooperation with government authorities, supra-regional institutions (such as higher educational institutions), and neighbouring local authorities. Persuasive Leitbilder can be helpful in this context.

Another function is motivating the population. An interesting form of participation is certainly the resort corporation model in Bad Schlema, where almost 200 private persons are shareholders of the municipal resort company, thus making an essential contribution to the development of the health spa at their own, private risk. In Gräfenhainichen, overall concepts were communicated from the outset in the media, so that, for instance, the excavator city Ferropolis gradually became a symbol with which the population could identify. Former miners proud of their heritage also identify with it. These examples show that Leitbilder can generate enthusiasm among many important actors and are not mere decoration for local development processes.

Finally, over and above the coordination and motivation functions performed by Leitbilder, their image-enhancement and marketing functions should not be underestimated. The addressees are local residents (internal marketing) and potential visitors, health resort patients, and investors (external marketing). In view of the decline in population registered by all the towns under study, it is particularly important to send positive signals influencing people in their choice of a place to live and persuading investors in search of promising ventures. The three towns under study have sought to achieve this, drawing attention to themselves with mottos like "Naturally Schlema," "Town with New Energy" and "Sport, Recreation, Health Region." The Bad Schlema and Gräfenhainichen Leitbilder, in particular, promoted the communities' unique characteristics. Such specifics were lacking in Johanngeorgenstadt. The Leitbilder for the small towns subject to this study thus fulfilled multiple functions:

- The orientation function was "omnipresent" and is especially important in imparting a sense of purpose in periods of radical change.
- The coordination of actors was particularly important in collaborating with external actors at superordinate levels of action and in enlarging the capacity for action.
- In motivating the population, "attractive" Leitbilder attain considerable reach.
- In the internal and external perception of the town (marketing function), Leitbilder stress unique selling propositions and call attention to the town.

General Role of Leitbilder

Finally, we review the importance of Leitbilder in each of the researched cases. Without a doubt, Bad Schlema pursued the straightest line of development. The idea of reconstructing the health spa was born, investigated, raised to the level of a Leitbild, and impressively implemented – apparently with consistency. Practically all endeavours and projects were directed towards attaining this goal. There was no alternative. The proof of the Leitbild has been in its success. Fifteen years on, Bad Schlema has changed beyond all recognition. Almost nothing remains to recall the "Valley of Death" where one "couldn't cross the street without rubber boots." But it is also interesting to consider what the result would have been if the concept had not worked out. Purely hypothetically, what would have happened if the radon source had suddenly dried up? Obviously, the Schlema Leitbild also entailed risks, for the success of a linear strategy approach is not automatic. Indeed, the development of Bad Schlema would have proved impossible if certain underlying conditions had not been met (comprehensive government remediation; "sponsorship" of the state government of Saxony for development of the health resort).

The example of Gräfenhainichen certainly best meets the demands made by a strategic planning concept embodied in a Leitbild approach. Planning was understood as an ongoing and flexible process. The lack of planning capacity in the town was compensated by strategic cooperation with creative, external experts. The Leitbild was developed and updated in an open process. Evaluation of the situation also played a role, enabling the main actors to reflect on what had been achieved. What the town could not manage on its own, it sought to implement through regional partnerships. Ferropolis provided a strong, supra-local symbol for renewal in the direction of tourism, culture, recreation, and quality of life on the basis of sustainable sources of energy. But in this case, too, it should be noted that the Leitbild could be implemented only with the strong support the municipality was provided with by comprehensive government assistance in mining remediation and by other initiatives and programmes.

Finally, the third example, Johanngeorgenstadt, shows how fraught with difficulty and how daunting the generation and implementation of Leitbilder can be. A succession of Leitbilder was drafted to guide the development of the town: reconstruction of the old town core, then alternative community development with several sub-centres, and finally the regional model. Ultimately, all attempts failed to get under way because the general conditions for successful implementation were lacking, because the force to translate the plans into action was wanting, or because it proved impossible to attract political attention. Neither internal nor external stimulus led to concerted action. The case impressively demonstrates that Leitbilder can be effective only if they are creatively integrated into a planning and development process, which has yet to be achieved to the necessary degree in Johanngeorgenstadt. The Leitbilder of the three towns under study can be summarized as follows:

- Bad Schlema: the Leitbild was an effective development goal with long-term stability; classic, linear understanding of planning; but capable of implementation only if all framework conditions are obtained in the long run.
- Gräfenhainichen: step-by-step, evolutionary formulation of the effective Leitbild, parallel implementation (open, flexible, adaptable); strategic approach; planning as a creative process undertaken with the aid of external expertise.

- Johanngeorgenstadt: the Leitbilder developed proved largely illusory; they gained neither persuasive power nor leverage and had little impact on the development process.

Regardless of the success enjoyed in the first two cases, it should not be overlooked that all three towns still have to cope with high unemployment and outmigration. Under the general conditions that the structural paradigm change in 1990 produced, even very good strategic concepts in conjunction with strong government support were unable to produce any final answer to the structural problems faced by small eastern German mining towns.

6.4 Conclusions

The aim of this study has been to examine how important Leitbilder have been as an element of strategic planning in handling structural change in small towns. A comparative case-study design was applied in examining three small mining towns in the eastern part of Germany that had to cope with extreme structural change following the unification of Germany.

The study demonstrates the great importance of Leitbilder for the development of the small towns concerned. After German unification in 1990, all three moved rapidly to re-orientate themselves, and finally developed comprehensive written Leitbilder. The functions of the Leitbilder – such as orientation, coordination, motivation, and marketing – were fulfilled. However, one case shows that the Leitbilder which were elaborated did not necessarily bring the desired success.

Although a rough division into three phases can be discerned (search, self-discovery, consolidation), a uniform pattern in the drafting and implementation of Leitbilder is, as expected, not apparent. Procedures, forms, and functions differed widely. For example, in one case a Leitbild agreed upon at the outset was implemented unswervingly and consistently. In another case, the Leitbild evolved over time and implementation ran parallel to the process. In all cases, however, ideas in the minds of the major actors sufficed to become effective Leitbilder. The coordinating function of the Leitbilder involved a high measure of cooperation with outside actors.

Six factors for successful Leitbild development and implementation can be identified. These serve as a basis for discussion whether the findings of the comparative study on the importance of Leitbilder for structural change can be generalised so they can be applied to all small towns. Moreover, the two final aspects show marked differences between small towns and cities.

Thesis 1 – Strength of the need for orientation: the greater the need for re-orientation in structural change, the greater will be the role played by Leitbilder. This thesis was a point of departure for selecting the case examples. It has been confirmed by the study. All things being equal, Leitbilder are likely to prove less important for small towns experiencing weaker structural change.

Thesis 2 – Availability of good starting situations: an attractive Leitbild is easier to elaborate if the town offers a favourable starting situation. In Bad Schlema, for example, the health spa resort concept was rooted in the history of the town, and in Gräfenhainichen there was the idea of Ferropolis. In towns that offer no such points of departure, Leitbilder are likely to play less of a role. However, it is also the job of the town to search for creative starting points.

Thesis 3 – Realism: The more realistic a Leitbild is, the easier it will be to implement. However, the degree of realism is very difficult to judge. Whether actors should rapidly renounce a certain Leitbild (flexibility) or should only persist more strongly in implementing it (perseverance) is one of the most difficult decisions for a community. Gräfenhainichen persevered successfully in developing Ferropolis and tourism, while Johanngeorgenstadt invested too much effort in ultimately vain concepts. Small towns with less realistic visions are therefore likely to enjoy less success with their Leitbilder.

Thesis 4 – Assertive power: The more assertive power a town has, the more successful any Leitbild will be. The power of a town to put its ideas into effect depends on its financial situation and its attractiveness for investors and the public. The degree of available government support also plays a major role. In all three cases under study, the financial situation was bad and attractiveness non-existent. In two cases there was considerable government support (government mining remediation). In the less successful Johanngeorgenstadt, there was little support for a substantial period. The majority of German small towns is likely to be in a much better starting position for development, but very few small towns could count on comparable support from the government in solving their problems.

Thesis 5 – Capacity: An attractive Leitbild can be more easily prepared and implemented if the town has adequate planning and control capacities. As we have seen in section 6.2, small towns are in a worse position in this regard than larger communities. However, the study shows in the case of Gräfenhainichen that it is possible to compensate for this disadvantage through openness and strategic contacts with institutes of higher education and planning firms. Small towns without such openness and strategic partnerships will tend to have a hard time with Leitbilder.

Thesis 6 – Manageability of the situation: Leitbilder are easier to develop, disseminate, and implement if the situation is easily comprehensible and manageable. The greater manageability of structures and actor networks in small towns as compared to large towns and cities (see section 6.2) provides a favourable setting for strategic action. Development processes can be clearly structured and the content limited to essentials. Under these circumstances, Leitbild and projects go hand in glove. Leitbilder are more concrete than is generally the case in cities. They are therefore much more plausible for addressees. As far as the transferability of the findings from the comparative case study to all small towns affected by structural change is concerned, manageability is likely to be similarly great in all small towns.

In all, the great importance of Leitbilder in the small eastern German mining towns revealed by the study can be said to hold true at least partly for all small towns experiencing structural change. At the same time, the study points out major differences in the framework conditions for developing and implementing Leitbilder in small towns and cities.

References

Adam B. (2005): Mittelstädte – eine stadtregionale Positionsbestimmung. Informationen zur Raumentwicklung, No. 8, 495-523.

Albrechts L. (2004): Strategic (spatial) planning re-examined. Environment and Planning B: Planning and Design, Vol. 31, 743-758.

Andersson M., Karlsson C. (2004): Regional Innovation Systems in Small & Medium-Sized Regions. A Critical Review & Assessment. CESIS Electronic Working Paper Series. Paper, No. 10, August 2004. Jönköping.

Aehnelt R., Kühn M., Schütte I. (2006): Lebensqualität in Klein- und Mittelstädten. Monitoring im Städtekranz Berlin-Brandenburg. REGIO transfer Vol. 6. Erkner b. Berlin: Institut für Regionalentwicklung und Strukturplanung (IRS).

Bataïni S.-H., Maillat D., Crevoisier O. (2002): A new approach to the attractiveness: The case of medium sized industrial cities. Jahrbuch für Regionalwissenschaft, Vol. 22, 41-59.

Baumgart S., Flacke J., Grüger C., Lütke P., Rüdiger A. (2004): Zusammenfassung der Diskussion und Ausblick, In: Baumgart S., Flacke J., Grüger C., Lütke P., Rüdiger A. (eds.), Klein- und Mittelstädte – Verkleinerte Blaupausen der Großstadt? SR-Papers Nr. 1. Dortmund: Universität Dortmund, 87-94.

Becker H., Jessen J., Sander R. (eds.) (1999): Ohne Leitbild? Städtebau in Deutschland und Europa, 2nd ed. Stuttgart: Krämer.

BMVBS (2006): Leitbilder und Handlungsstrategien für die Raumentwicklung in Deutschland. Berlin: Bundesministerium für Verkehr, Bau und Stadtentwicklung.

Courtney P., Errington A. (1999): Spatial patterns of economic activity: The role of small towns in the rural economy and some implications for planning and development policy. Contribution to RICS Research Conference – ROOTS '99. http://www.rics.org/Property/Ruralandnaturalassets/Ruralandnaturalassetdevelopment/spatial_patterns_of_economic_activity_19990101.html (retrieved on 22.10.2007)

Dehne P. (2005): Leitbilder in der räumlichen Planung. In: Akademie für Raumforschung und Landesplanung (ARL) (ed.), Handwörterbuch der Raumordnung. Hannover, 608-614.

Diller C. (2003): Regionalentwicklung durch neue Kooperationen – von wem und für wen? STANDORT – Zeitschrift für Angewandte Geographie, Vol. 27, No. 2, 79-84.

Dörner D. (2006): Die Logik des Misslingens. Strategisches Denken in komplexen Situationen. Hamburg: Rowohlt Taschenbuch Verlag.

Erickcek G. A., McKinney H. (2006): "Small Cities Blues:" Looking for Growth Factors in Small and Medium-Sized Cities. Economic Development Quarterly, Vol. 20, No. 3, 232-258.

Grabher G. (1993): Wachstums-Koalitionen und Verhinderungsallianzen. Entwicklungsimpulse und -blockierungen durch regionale Netzwerke. Informationen zur Raumentwicklung, No. 11, 749-758.

Healey P. (2004): The Treatment of Space and Place in the New Strategic Spatial Planning in Europe. International Journal of Urban and Regional Research, Vol 28, No.1, 45-67.

Hein E. (1998): Planungsformen und Planungsinhalte. In: Akademie für Raumforschung und Landesplanung (ARL) (ed.), Methoden und Instrumente räumlicher Planung. ARL, Hannover, 187-203.

Hutter G., Wiechmann T. (2005): Back to the future – Emergent Strategies in Strategic Spatial Planning. Conference paper. Aalborg: Regional Studies Association. 28.-31.05.2005.

Jessen J. (2005): Leitbilder der Stadtentwicklung. In: Akademie für Raumforschung und Landesplanung (ed.), Handwörterbuch der Raumordnung. Hannover: ARL, 602-608.

Knieling J. (2006): Leitbilder und strategische Raumentwicklung. Planungstheoretische Einordnung und Diskussion der neuen Leitbilder für die deutsche Raumentwicklung. Raumforschung und Raumordnung, Vol. 6, No. 64, 473-485.

Koch T. (2006): Stadtmarketing. Praxishandbuch für kommunales Management. Saarbrücken: VDM Verlag Dr. Müller.

Leimbrock H. (2000): Eine Stahlstadt im Strukturwandel: Die sächsische Stahlstadt Riesa auf dem Weg zu einer neuen Identität. In: Danielzyk R. et al. (eds.), Sanierung und Entwicklung in Ostdeutschland – regionale Strategien auf dem Prüfstand. IÖR-Schriften Vol 32. Dresden: IÖR, 113-128.

Leimbrock H., Lintz G. (2002): Environmentally focused cooperation projects as a stimulus for the development of old industrialised regions. Case studies in eastern German regions in which small and medium-sized towns predominate. In: ERSA (ed.) From industry to advanced services - perspectives of European metropolitan regions, 42nd Congress of the European Regional Science Association held in Dortmund 2002. Book of abstracts. Dortmund, 16.

Lindblom C. E. (1959): The Science of "Muddling Through". Public Administration Review, No. 19, 79-88.

Lutter H. (2006): Neue Leitbilder der Raumentwicklung in Deutschland. Raumforschung und Raumordnung, Vol. 6, No. 64, 441-450.

ÖIR (2006): The Role of Small and Medium-Sized Towns (SMESTO). Final Report, ESPON 1.4.1. Wien: Österreichisches Institut für Raumplanung.

Rüdiger A. (2004): Die Aufgabe der Selbstverwaltung in Klein- und Mittelstädten. In: Baumgart S. et al. (eds.), Klein- und Mittelstädte – Verkleinerte Blaupausen der Großstadt? SR-Papers No. 1. Dortmund: Universität Dortmund, 41-46.

Sartorio F. S. (2005): Strategic Spacial Planning. A Historical Review of Approaches, its Recent Revival, and an Overview of the State of Art in Italy. disp, Vol. 3, 26-40.

Schäuble I. (2004): Soziale Stadt in kleinen und mittleren Kommunen. In: Kurzprotokoll des SRL-Arbeitskreises „Soziale Stadt/Präventive Stadtentwicklung“ Short minutes of the SRL working group „social city/preventive urban development“. Kassel, 2-4 http://www.srl.de/dateien/dokumente/de/ak_soziale_stadt_protokoll_sitzung_7_2004-11-13_kassel.pdf

Sieverts T. (1999): Was leisten städtebauliche Leitbilder? In: Becker H., Jessen J., Sander R. (eds.), Ohne Leitbild? Städtebau in Deutschland und Europa. Stuttgart: Krämer, 21-40.

Wirth P., Müller B., Rathmann J. (2003): Rehabilitation and development area: uranium mining in south-west Saxony, Shaping structural change through co-operation. In: Klapperich H. et al. (ed.), ECI International Conference on Green Brownfields II from Cleanup to Redevelopment, Schloss Pillnitz, Dresden, Germany, 15.-19.06.2003, . Verlag Glückauf: Essen, 253-263.

Yin R. K. (2002): Case Study Research, Design and Methods, 3rd edition. London, New Delhi: Thousand Oaks.

7 Urban Metabolism and the Surface of the City

Clemens Deilmann

Introduction

Building, maintaining, and changing the urban fabric not only consumes and accumulates great quantities of material and energy resources, but also requires large capital investments. The biggest share of primary energy consumption is caused by building use and access (traffic). In Germany, building activities account for a 10% share of annual total gross domestic product, the flow. The net noncurrent asset represented by residential buildings, the stock, holds 50% of all noncurrent assets, and the non-residential building stock accounts for an additional 35%. Furthermore, the "urban fabric" is itself a source and prerequisite for human activities and progress of growingly urbanized societies. The competitiveness of societies will depend on how efficient the stocks and flows of the urban fabric are managed and on the ability to adapt the urban fabric to change. Looking at the material flow from an environmental point of view, limitations appear at the front and back end of the system: "While on the input side, the availability of resources is hampered by growing global competition for inexpensive materials, on the output side, the potential for dilution in water, air, and soil is decreasing with each load of pollutants. Because for many materials, anthropogenic flow rates have already passed geogenic flow rates, the disturbance of environmental compartments is at stake as exemplified by the fossil fuel driven climate change" (URBANMET 2007). Hence, resource consumption as well as environmental protection – the consideration of sources and sinks – need to be linked with each other and investigated as a whole.

The framing conditions of urban systems often alter rapidly – in terms of the long lasting life of the elements of urban fabric: Changes in technology, economy, population dynamics, environment (especially climate), and shortages of specific materials and energy sources require that the urban fabric is adapted to these developments. By modelling the urban development and thinking in scenarios 30 to 50 years ahead, it is possible to foresee constraints and challenges. The concept of urban metabolism is a powerful tool to describe, analyse, and model the stocks and flows of urban systems. This paper focuses on the urban fabric, "the hardware" of an urban system. It discusses how to examine the urban fabric in terms of form, age and typical structures, technical and non-technical surfaces, land use, spatial location, material composition, and consumption of resources.

The first chapter will specify the urban metabolism and the state of the art. The second chapter presents advances on the methodology by use of the urban structural type approach. Third, the eastern German case study scenarios will illustrate indicators and findings generated by the new metabolism approach. The paper will conclude with some challenges and the key issues of urban analyses.

Up until now, material flows and related matter are not on town-planners' minds. Therefore it will be necessary to use the results of Urban Material Flow Analyses as an add-on to the general discussion about land use and energy issues of urban planning, which are quite well perceived. In general, Material Flow Analyses (MFA) and Energy Analyses (EA) will strengthen the orientation of planners towards qualified urban density and deliver a whole new set of arguments for the renaissance of towns and resource-efficient settlement structures.

7.1 Definitions and State of the Art[1]

The metabolism of cities can be characterised by two different categories: the rapid throughput of materials or assets such as water, air, fuel, and most consumer goods, and the longer term assets such as buildings, networks, and infrastructure called urban hardware or the "urban fabric". The urban fabric holds a very large potential from the point of view of resource conservation, material recycling, and environmental protection, and to separate urban fabric completely from the remaining metabolic processes is inappropriate. When focusing on the fabric it is necessary to at least reflect those issues that are closely associated with it, such as energy use for heating and lighting (directly related to buildings design) or management of urban surface water as an example.

The urban fabric is defined as the aggregated stocks and structures of built environment in a city, including buildings (residential, commercial and industrial), transportation infrastructure (roads and rails) and technical infrastructure (energy supply, water distribution and collection networks).

Urban metabolism analyses so far could rely on the assumption that cities would grow and accumulate materials so that stocks permanently increase. European countries have experienced high economic growth since World War 2, and as a consequence of this the various urban fabric stocks have increased significantly. A recent study by Moll et al. (2003 and 2006) reveals that, despite large differences between countries, there is an unsustainably high level of resource use in European countries. In the previous EU-15 economy, the average direct material input had been nearly constant since 1980: about 16.5 tonnes per capita annually, with about 10 tonnes per capita per year accumulation to material stocks, out of which the largest part terminated in urban fabric stocks, and where construction minerals were used for the construction and maintenance of buildings (about 40 %) and for infrastructures (about 60 %). These stocks can be seen both as a future source of valuable materials as well as a future hazard for the environment. Therefore it is

1 Part of the text chapter 1 is based on contributions from Helge Brattebo, Stefan Bringezu and Paul Brunner, partners in the joint EU-Proposal URBANMET (2007).

very important to know the growth and the dynamic of change.

According to Müller (2006), who studied the dynamic behaviour of the Dutch residential system, material stocks, resource demand, and waste generation are very much influenced by three determinants: diffusion of service unit stocks, diffusion of materials used to produce these service units, and the lifetime distribution of ageing stocks. The results indicate the following: construction and demolition flows follow a cyclical behaviour; the cycles of construction and demolition flows are phase-displaced in the first half of the 21st century, with decreasing construction and increasing demolition; and growth of the dwelling stock is becoming increasingly more material intensive as the share of replacements is increasing.

Bergsdal et al. (2007), Bohne et al. (2007), and Holck-Steen et al. (2007) made use of dynamic MFA techniques to examine the long term behaviour of the Norwegian housing stock and the associated material and energy flows, waste generation, and environmental and economic impacts. The results confirm Müller's conclusion that there is indeed a strong dynamic and non-linear relationship between stocks and input and output flows. Future decades will be characterised by growing stocks, temporarily reduced (construction) inflows and strongly increasing (construction and demolition waste) outflows. There are significant potentials for material substitutions, and different materials have significantly different environmental and economic impacts. Future design, renovation, and management strategies will indeed play an important role in order to reduce negative impacts. The functional lifetimes of the built environment are very important in this respect.

The theoretical foundation for urban metabolism studies was laid down by Baccini and Brunner with "The metabolism of the anthroposphere" (Baccini and Brunner 1991). The above mentioned material flow accounting – Moll, Bringezu, Schütz (2005) on the level of national economy, Müller (2006) or Buchert et al. (2003) for the housing stock – delivers the quantitative background for more spatially differentiated studies on the metabolism of urban regions.

Existing urban metabolism studies, for example Vienna (Daxbeck et al. 1996 and 2001) or Toronto (Kennedy 2003), are generally concerned with capturing the most complete picture of all fluxes of energy, water, materials and wastes going in and out of an urban region. They cover urban fabric, consumer goods, services and throughputs like water, air and energy. Due to the thematic complexity of these analyses, they avoid further spatial differentiation into small areas like neighbourhoods; but when one focuses on the urban fabric and the dynamic of its change, analysis down to a micro scale becomes inevitable.

At the level of single constructions, knowledge and data exists on material flows and stocks, energy consumption, and life-cycle-wide environmental impacts for selected types of buildings, mainly for private homes[2] (Deilmann 2001). The

2 e.g. The BASiS IT-tool for analysis of material flows for 'Construction and Housing' for modelling of the development of the housing stock in Germany, over time and under different scenarios, and for calculation of material flows and environmental impacts: http://www.oeko.de/capabilities_services/own_analysis_tools/basis/dok/148.php

housing sector has been thoroughly discussed (see Lowe 2001; Rovers 2004; Kohler et al. 1999a), while interest in the technical systems for transport and utilities has so far been comparatively low. However, a cursory glance at material flows in the construction industry indicates that the areas of roads and utility infrastructure should be accorded a greater importance in the framework of a sustainable and environmentally friendly housing development (OECD 2005; Schiller 2007).

In addition – since the European "house" is almost erected – sustainable urban fabric stock management and operation and maintenance strategies are gaining increasing importance. There is also the ageing phenomenon related to such stocks: as they become old, more renovation and refurbishing investments are needed, as well as demolition and replacement by new construction inputs. Furthermore, the internal patchwork of growing and shrinking neighbourhoods, or up- and downgrading, which can be found in all cities, can only be grasped by micro-scale modelling. The latter is of even greater importance when the population of cities does not grow, but stagnates, and therefore redistribution, restructuring, and internal migration are of prior interest. In cities with constant decrease in population density, concerns about infrastructure issues and maintenance are inevitably growing, as less people have to shoulder the efforts and costs.

These are the reasons why research of the Leibniz Institute of Ecological and Regional Development (IOER) became focused not only on sustainable use of natural resources, but also on questions concerning the use of existing "built" resources without loss of efficiency. Only a highly spatially differentiated analysis can contribute to answer these questions and serve as basis for a monitoring of the dynamic of stocks and flows related to the internal reconstruction of the urban fabric. In order to do so, IOER has developed an UST approach which makes use of advanced GIS-tools and which can be integrated into the concept of urban metabolism.

7.2 Urban Structural Type Approach and Urban Metabolism

A practical and appropriate way to describe specific properties of the urban fabric of a city is by use of typologies for buildings and urban areas with characteristic layout of buildings and open spaces. Typologies in spatial science are the essential way to reduce complexity, and recognise, discover, and analyse spatial development phenomena. Typology approaches are used to describe buildings, infrastructure, and the urban structure. USTs are basic spatial units with physiognomic homogenous characteristics, designated by characteristic formations of buildings and open space such as development patterns like neighbourhoods dominated by single-detached family homes or linear developments of the 1950s. They describe land use areas with similar environmental and infrastructure conditions and similar use (functions) and indicate the time of origin. They provide a picture of the morphological situation of the urban area (Duhme and Pauleit 1999) while at the same time representing core characteristics for structural analyses, synthesised from representative objects and databases. Basic sets of pre-defined USTs exist in literature, but the USTs used for analyses will usually have to be adapted according to the specific objectives of the investigation, for example if environmental issues of green areas or building material issues need to be covered.

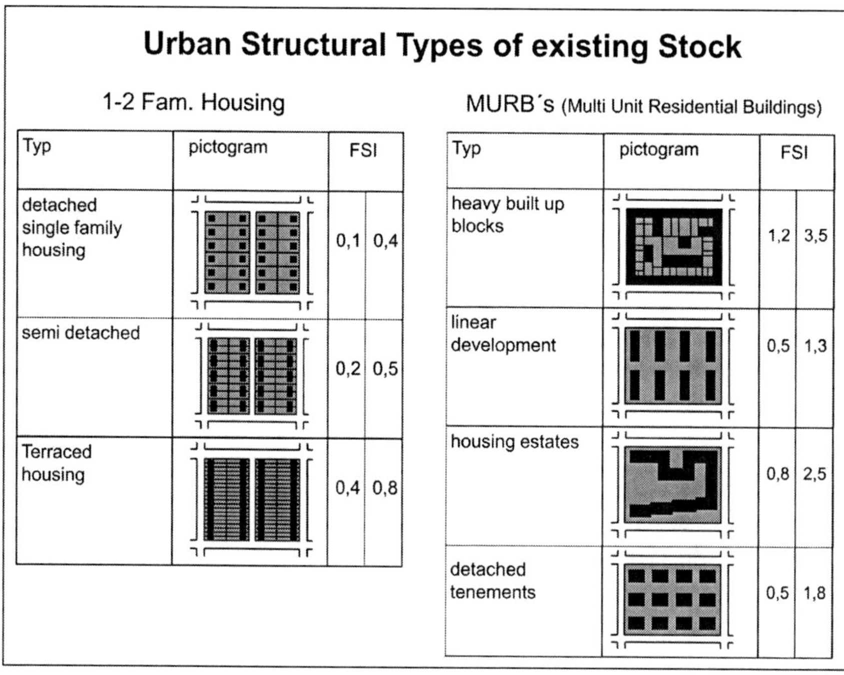

Fig. 1: An example of Urban Structural Types for residential buildings.

Each UST is characterised by typical urban density values and, as result, USTs as spatial units interlink buildings and related infrastructure. Small-scale parameters for infrastructure equipment, e.g. length and profile of pipes, as well as population density parameters are both based on empirical and statistical data that can be attributed to USTs. This is an important and necessary step in order to illustrate the effects of urban density on infrastructure elements (Schiller 2007). The resulting "city map" allows calculation of resource intensity and expenditure at city level, for districts, neighbourhoods, or micro-scale (1-5 hectares).

By use of typology and related databases like GEMIS or ECOINVENT[3] it is possible to calculate stocks and flows. It is also possible to calculate the city surface, which can be understood as the interface between the man-made artefact and the

3 www.ecoinvent.org and www.gemis.de

environment. The city surface comprises municipal roads, squares, private access ways, building surfaces such as external walls, windows, and roofs, and non-built spaces such as urban green parks and private gardens. The relevance of this interface is exemplified most evidently by the "thermal envelope" which determines the energy consumption (transmission heat) of buildings and is exemplified by the facade as the "source" of the rainwater runoff from building materials. By modelling the city surface it is possible to make assumptions about the maintenance necessities of these surfaces and the technical infrastructure below ground, as well as performance indicators regarding ecological functions of sealed and unsealed surfaces.

The empirical recognition of urban types can be done visually by hand on the basis of maps and aerial photography or with the support of GIS tools. The state of the art depends on the geo-data availability for the cities. Some cities have a digital building mask of the entire administrative area, and the vector-data provided thus allows almost one-to-one calculations for the built-up area. The polygons of buildings and land area are attributed with certain information such as age, size, number of dwellings, population density etc., which simplifies the recognition of building types and urban structures.

Another option - if cities don't provide this information - is automatic settlement recognition and characterisation on the basis of remote sensing data (Michel et al. 2006) or through urban structural type recognition and evaluation based on digitised topographical maps (Meinel 2006). This recognition proves to be successful with a 1:25,000 scale and bigger. IOER developed the ArcMap application add-on SEMENTA® (**S**ettle**ment A**nalyzer) on this basis. This tool makes use of topographical map DTK25(-V) and the ATKIS Basis-DLM to give a fully automatic analysis of settlement planning parameters. Complex image processing software is used to extract, measure, and classify the total housing stock from the object outline layer in DTK25(-V). ATKIS block geometry can then be applied to calculate settlement structural parameters such as building type, building footprint, floor space, building volumes, or the number of flats and occupancy for individual blocks. The program was developed to offer, for the first time, an objective and efficient investigation of parameters for settlement structures over large areas and with high spatial resolution.

The use of advanced GIS-tools will support the analyses of urban areas, since typical buildings and infrastructure elements can be precisely detected in a given urban settlement area (Thinh 2005). Material flows and energy analyses and stocks can be *spatially* differentiated for different scales of investigation. A well differentiated picture of the urban fabric and city surface allows extensive monitoring of the urban metabolism down to the "micro scale" of neighbourhoods, which is a necessary step to understand the in-, out- and throughput of materials caused by the permanent reconstruction and extension of the urban fabric.

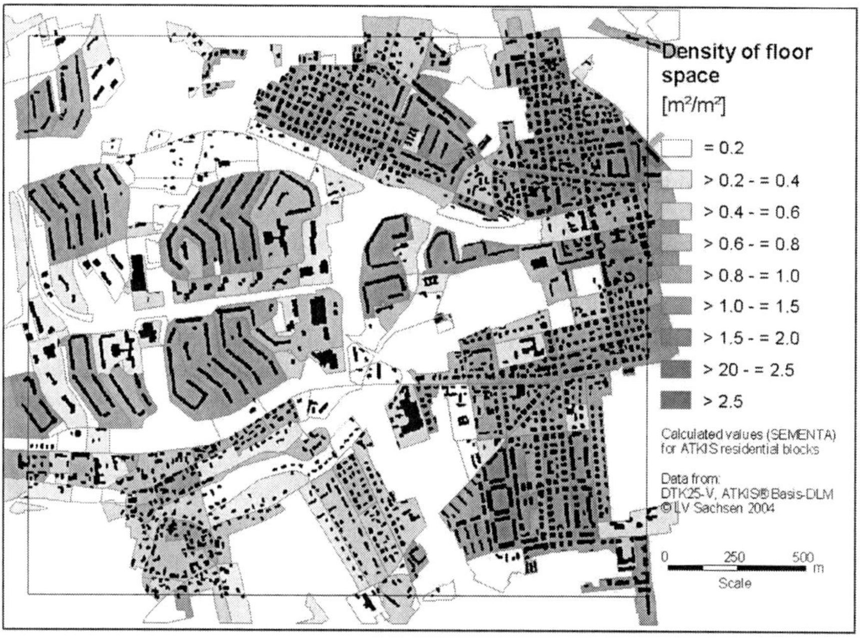

Fig. 2: Example results for floor space density, map of Dresden-Gorbitz.

7.3 Eastern German Case Study Scenarios – the Surface of the City 2020

The eastern German case of cities with population decline of 1% per year over the last decade has drawn a lot of attention, and it advances to a living lab for driving factors and effects of shrinkage. Unfortunately the term "shrinkage" gives the wrong impression. Opposite as the word infers, the cities are not getting smaller in physical terms, but they are characterised by a process that may be denoted as one of "extensivation" or "perforation": The material stock might decrease but the surface of the city often keeps growing.

The investigation presented here was carried out as an in-depth case study of three eastern German Cities. Focus of the investigation was the residential building stock and the technical infrastructure relevant for the accessibility within the residential area. Main roads for transit or non-residential purposes are excluded. The research findings will be presented via a fictitious synthesised city of 250,000 inhabitants and an expected decrease in population of 15% until 2020. The vehicle of a fictitious city was necessary, because involved cities asked for anonymity.

The key methodology was as described with the metabolism concept in conjunction with the UST approach. Furthermore, a scenario-writing technique was adopted. The scenario settings were discussed with experts and agreed on with representatives of town planning departments. As the expected housing demand can be realised by very different development paths of land use and resource consumption, it is important to discuss the interdependence of housing loss (demolition of dwellings, increase of vacancies) and housing production (new building, extensions/refurbishments), as well as the use of brownfield versus greenfield development, together with the planning authorities.

Based on this empirical exercise, the housing demand, residential USTs, and building types and age groups for the city were analysed ex-post from 1995 to 2001 and then predicted for 2020. Within the housing stock in 2001 more than 25% of the dwellings were vacant. This was the starting position. The further decrease in housing demand due to population loss of 15% until 2020 cannot be outweighed by the increase of demand due to the reduction in size of households (from 2,01 to 1,87 persons/household). Demolition of dwellings is necessary to ensure economic stability of the housing market. Two types of multi-family houses - old Wilhelminian style buildings from 1918 and earlier and prefabricated buildings from 1971 to 1990 - are especially affected. In contrast to these dwellings the stock of single-family houses will grow further.

These were the framing conditions for two scenarios. The trend and the regeneration scenarios were defined in cooperation with the town planning experts in order to illustrate the change of stock: The trend scenario reflects the most probable development. For new construction it is assumed that 20% will be built on brownfield sites, 30% on empty plots and 50% on new land (greenfield sites). The demolition activity amounts to approximately 1% of the residential building stock from 2001 per year. Till 2020 the vacancy will thus be reduced by 5% to a level of 20%.

In the regeneration scenario the decrease in vacancy is clearly stronger. It will decrease until 2020 by around 12% to a level of 13%. The demolition rate amounts to 1.3% per year of the residential building stock from 2001. The assumptions for new buildings are defined slightly differently. In this context, the use of brownfield sites and activities for consolidating existing urban structures have a high priority.

For all the following findings, the data were aggregated and the figures comprise averages or sums for the entire city. However, the basis for these figures is the micro-scale modelling of urban areas along USTs. The city development patchwork made up of areas with new construction or those areas with demolition or reconstruction of buildings is therefore contained in these figures and may be used as basis for further analyses.

The Surface of the City

The following results are based on geometrical information attributed to building types and USTs. The quantities are calculated according to architectural arithmetic volume/surface rules for the scale 1:100.

The ex-post analysis for the year 2001 (left side of figure 4) illustrates at first that the typical eastern German residential area comprises on average 105m^2 of technical surface (external walls, windows, roofs, streets) per inhabitant and roughly the same

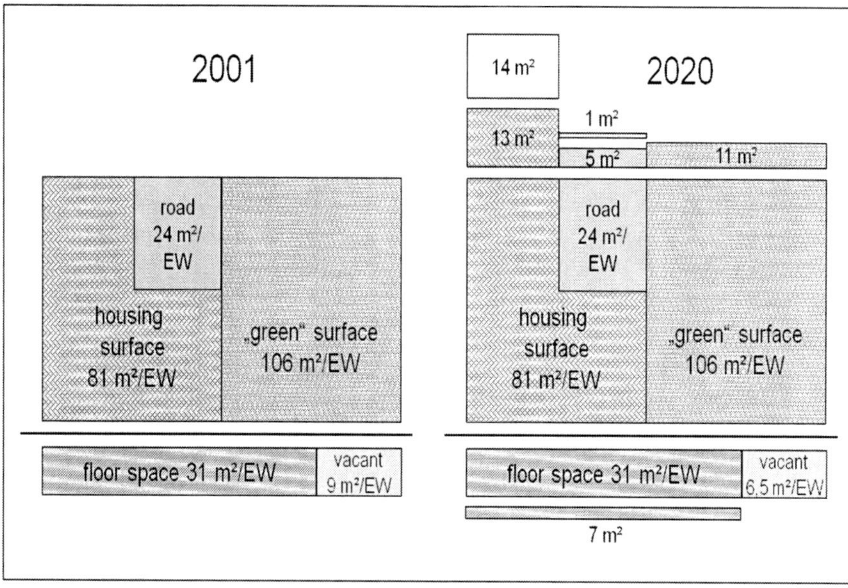

Fig. 3: Surface consumption per inhabitant 2001 and 2020.

amount of green area for the sake of 31 m² net floor space living area per inhabitant. Thus, the ratio between surface and floor space is 7 to 1. This ratio varies between the different USTs, for example multifamily housing and single family home areas, from 3:1 to 9:1.

The trend scenario for 2020 shows the growth of surface per inhabitant due to future consumption patterns, i.e. bigger flats and more single family homes, and all in all the additional space and surface consumption per citizen will grow by about 20% in both scenarios until 2020. Within this increase, it is of high relevance that the technical infrastructure, roads and mains, extends while at the same time the population declines, meaning that fewer citizens will have to shoulder the maintenance of rising areas and costs of infrastructure. 1% less density causes 1% more infrastructure cost (Siedentop et al. 2006). In the fictitious city, the density will drop by 20% in both scenarios. At present such prospects trigger off a vehement debate in Germany about who should pay the rising costs of infrastructure and whether or not there is a chance of implementing the "consumer pays" principle for technical infrastructure by taking different levels of urban density into account. The concept of urban metabolism can assist in explicitly and concretely defining the quantities to be expected.

Comparing the scenarios and looking at the development of the shrinking cities in absolute terms (millions of m² of city surface) shows a stable, slightly growing area for infrastructure and green area in both scenarios (figure 5 Trend and Regeneration). Residential floor space decreases by 2-3% and the building surface by 2-8%, given a population decline of 15% and a successful demolition program. If the demolition activities don't show the desired result and possibly remain around the long-term average of 0.2 – 0.3% of stock per year, the building surface would even grow slightly. The development seen in relative terms (m² per inhabitant) presents a picture of growing surfaces in all categories (figure 6 Trend and Regeneration). All these surfaces need to be maintained and publicly or privately paid for. Most of the burden has to be carried by the consumers themselves. But looking at the infrastructure area, it is the responsibility of the municipality to guarantee access and a supply of electricity, water, gas, and sewage for the citizens, and the growing efforts will have to be paid by tax payers (for urban green and roads) and the consumers (for most of the mains).

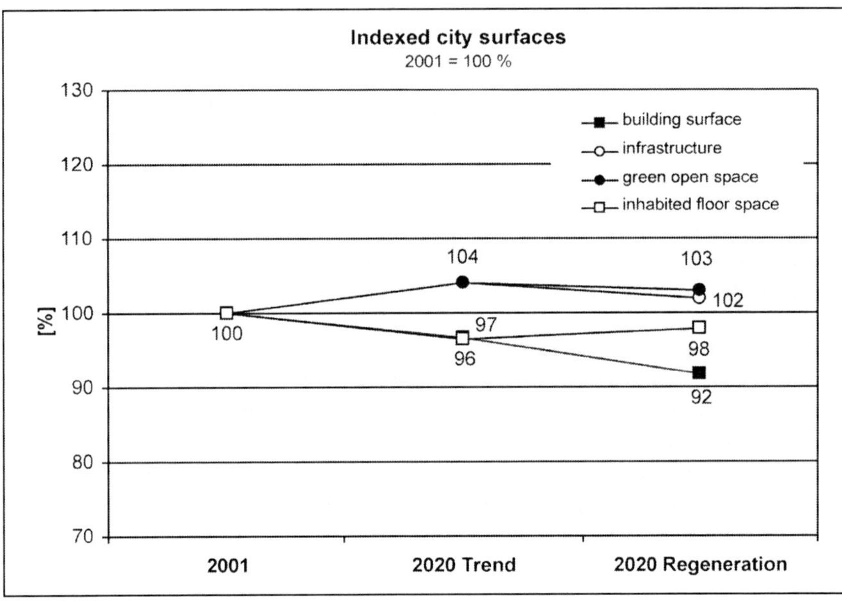

Fig. 4: Indexed city surfaces.

Further Effects of the Extensivation of Cities in Brief

Vacancy in the housing stock – that may reach up to 30% in eastern German neighbourhoods – enlarges the thermal envelope of buildings in relation to the inhabited area. When the housing volume in use looks like "Swiss cheese", there will be an additional heating demand due to the "passively" tempered vacant dwellings. Calculations show that 20% of vacancy generates an additional 4% of heating energy demand for the inhabited dwellings (Deilmann 2005). If one assumes that annually 1% of the building stock in the city will be thermally improved to reach a 40% heating demand reduction for these refurbished houses, the CO_2-emission per average inhabitant of the city won't nevertheless decrease until 2020, due to high vacancy rates and increased floor space consumption per inhabitant.

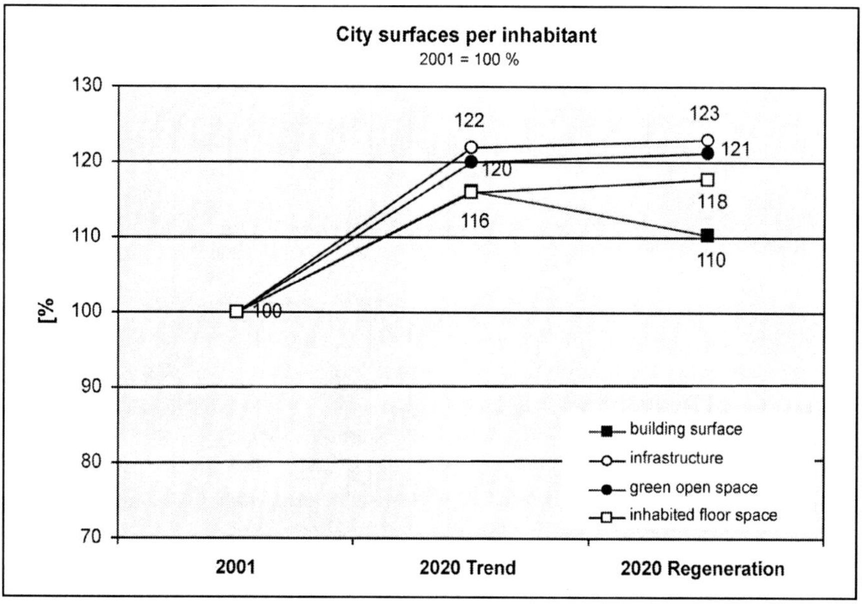

Fig. 5: City surfaces per inhabitant.

Concerning demolition waste quantities, this depends on the "desired" demolition rate, for example with respect to market stabilisation. The material composition of demolition waste was calculated for the two scenarios mentioned above, according to the settings given by the planning authorities (defined areas of city, more or less prefabricated housing versus more or less Wilhelminian style houses, etc.). In the result, the quantity and the kind of the demolition waste was computed. In the regeneration scenario, almost 1000 tons of demolition waste per day must be transported out of the city. In the trend scenario, it would still be 700 tons per day. In comparison, the material consumption for new residential construction is far lower. In both scenarios the material flow out of the city is 2.5 times bigger than the material inflow. The housing stock of the city is no longer an accumulator of materials, but demolition activities will make the city a net "material" supplier (Deilmann and Gruhler 2005).

Nevertheless and despite population decline, new building construction will continue on a low level. With respect to the development of the city surface and especially the infrastructure, it is decisive *where* the new construction will happen: brownfield sites, empty but developed plots, new land etc. There is a great potential in using the new brownfields produced by demolition of buildings. In the case of the trend scenario, about 20 hectares per year will become available – for example for inner city single family home developments. In the regeneration scenario there are even 30 hectares available per year, as the demolition activity is higher. Although both scenarios assume a high proportion of brownfield re-development, the land "output" by demolition will still be higher than the demand calculated for new building activities. In the trend scenario, the difference amounts to 7 hectares per year, which means a permanent growth of brownfield areas. In 2020 this will add up to 130 hectares for the trend and 150 hectares for the regeneration scenario.

7.4 Conclusions

The physical characteristics of urban areas can be analysed and modelled on a micro-scale by typologies of buildings and USTs. The spatial analyses can be supported well by GIS-tools for automatic feature (such as UST) recognition. Digital geo-information linked with statistical data can reach high congruency to on-site empirical analyses, even so for some socio-economic indicators like population density, floor space per inhabitant etc. A typology of buildings and urban structure can be linked to environmental databases for products and construction elements in order to generate the quantitative backbone of resource use. The analyses bring up a whole sequence of characteristic features and indicators for defined urban areas or entire cities. Main intention of IOER research in the context of urban metabolism is to link MFA and EA with spatial information to provide spatially differentiated knowledge for more sustainable urban development. With this intention, urban metabolism research orientates not only on the status quo, but investigates the dynamic of change and uses scenarios for ex-ante studies. By setting instructive scenarios in cooperation with planning authorities, technical experts, and research partners, it is possible to illustrate effects of different development paths and raise the awareness of urban administration and planning, as well as of politicians towards the management of natural resources and for the efficient use of existing built resources. This is the challenge.

In general, the predominant concern within urban planning is land use and allocation of functions. City planners and managers are widely lacking precise knowledge about the material and energy flows and the material stocks related to and triggered by the built environment, i.e. the buildings and infrastructure systems, the different urban forms, and management practices determining the impacts associated with construction, use, repair, renovation, recycling, and disposal. But as it stands, the urban design defines the types of buildings and infrastructure, and thus, depending on regional characteristics and material preferences, determines a significant share of overall resource consumption of the urban fabric and therefore of the overall resource efficiency of the urban system.

In addition, urban planning, which deals with small scale regeneration, infill, green or brownfield development in relation to intra-urban migration, often tends to consider the overall effect in resource terms and impact as low, not traceable, and anyhow demand driven. The internal patchwork of growing and shrinking neighbourhoods, or up- and downgrading, which can be found in all cities, is difficult to grasp in resource terms and in any case needs micro-scale modelling. The latter is of even greater importance when the population of cities does not grow, but stagnates, and therefore the redistribution, restructuring, and internal migration are of prior interest. In this case, the question is how to use existing "built" resources without high loss of efficiency.

The growing urban fabric stocks and their changing characteristics as well as design and management challenges most likely will continue to gain importance with respect to the sustainability and competitiveness of European cities. It is necessary to have instruments at hand to anticipate the consequences of such changes for the urban metabolism, and to have reliable tools and information to support decisions towards resource optimization of the urban metabolism. Improvements towards a more sustainable city and visions (e.g. zero emission cities) need to be reflected in terms of the urban metabolism. The methodology presented can produce highly aggregated pictures of the urban built environment for a national economy or for an entire city, as well as micro-scale pictures of stocks and flows on neighbourhood level or for housing market segments. It is also able to illustrate short and long term effects. This allows the consideration of metabolic constraints and requirements in the planning of the urban fabric, providing concrete guidance and reference tools for planners and engineers considering the design and maintenance patterns of buildings and infrastructure in the context of sustainable urban form.

7.5 Acknowledgement

The research on energy and mass flow in urban areas was mainly funded by SMWK (Saxony Ministry of Science and Fine Arts) (NAWO 2001) and UBA (German Federal Ministry for Environment) (Buchert et al. 2004). A special thank you to all participants and staff members of IOER Dresden and to Prof. Herz from Dresden University of Technology (TUD), Chair of Urban Engineering. Thanks also for the input from our partners of the EU-proposal URBANMET, Prof. Paul Brunner, Vienna University of Technology (TU Wien), Prof. Stefan Bringezu, Wuppertal Institute for Climate Environment, Energy (WI) and Prof. Helge Brattebo from the Norwegian University of Science and Technology (NTNU).

References

Automata – A Case Study of the Dresden City Region since 1780. In: Filho L.; Gomez, M.; C. Rautenstrauch (eds.), Second International ICSC Symposium on Information Technologies in Environmental Engineering, Proceedings. Aachen, 349-364.

Baccini P. and Brunner P. H. (1991): Metabolism of the Anthroposphere. New York: Springer.

Bergsdal H., Brattebø H., Bohne R. A., Müller D. B. (2007): Dynamic Material Flow Analysis for Norway's Dwelling Stock. Building Research & Information (accepted for publication).

Bohne R. A., Bergsdal H., Brattebø H. (2007): Dynamic eco-efficiency projections for C&D waste recycling strategies at city level. Journal of Industrial Ecology, 11(3) (in press).

Bringezu S., Moriguchi Y. (2002): Material Flow Analysis. In: R. U. Ayres and L. Ayres (eds.).

Brunner P. H. (2007): URBANMET – Urban fabric towards sustainable urban metabolism. Proposal part B to Work programme ENV.2007.2.1.5.1, FP76-ENV-2007-1.

Buchert M. et al. (2003): Sustainable building and housing in Germany: Material-flow-related components for a national sustainable development strategy ‚linking the building and housing sector with the complementary area of ‚public infrastructure' - Final report UFOPLAN-No. 298 92 303/02, in cooperation with IOER and TU Dresden, Darmstadt/Dresden, July 2003 (publication in the ‚Texte' series of the Federal Environmental Agency (UBA). UBA (2004), Nachhaltiges Bauen und Wohnen in Deutschland, UBA-Texte, No. 01, Berlin).

Daxbeck H., Lampert C., Morf L., Obernosterer R., Rechenberger H., Reiner I., Brunner P. H. (1996): Der anthropogene Stoffhaushalt der Stadt Wien. (Projekt PILOT). Technische Universität Wien. Institut für Wassergüte und Abfallwirtschaft. Abteilung Abfallwirtschaft. Wien.

Daxbeck H., Kilialova A., Obernosterer R. (2001): Der ökologische Fußabdruck der Stadt Wien, Ressourcen Management Agentur (RMA). Final report, commissioned by Magistratabteilung 22 – Umweltschutz der Stadt Wien. Wien.

Deilmann C. et al. (2005): Kommunales Stoffstrommodell Wohnen – Szenarien 2020 für den Wohnungsbestand am Beispiel zweier disparater kommunaler Entwicklungen. Final report. Dresden: IOER.

Deilmann C., Gruhler K., Böhm R. (2005): Stadtumbau und Leerstandsentwicklung aus ökologischer Sicht. München: Oekom Verlag, 103.

Deilmann C., Gruhler K. (2005): Stoff- und Energieflüsse von Gebäuden und Infrastrukturen als Grundlage für ein vorausschauendes szenariogeleitetes Stoffstrommanagement. Österreichische Wasser- und Abfallwirtschaft, No. 7-8, 103-109.

Deilmann C. et al. (2001): Nachhaltige Entwicklung des Wohnungsbestandes in sächsischen Groß- und Mittelstädten: Entwicklungsszenarien ausgewählter Wohngebiete unter ressourcen- und nutzungsorientierten Aspekten (NAWO). Final report.
http://www.ioer.de/Nawo

Duhme F. and Pauleit St. (1999): Stadtstrukturtypen – Bestimmung der Umweltleistungen von Stadtstrukturtypen für die Stadtplanung. In: Raumplanung, No. 4, 33-44.

European Topic Centre on Resource and Waste Management (2005): Environmental Input-Output Analyses based on NAMEA data. A comparative European study on environmental pressures arising from consumption and production patterns. Copenhagen: GOV/TDPC/URB, 1.

Gruhler K. et al. (2002): Stofflich-energetische Gebäudesteckbriefe – Gebäudevergleiche und Hochrechnungen für Bebauungsstrukturen, IÖR-Schriften No. 38. Dresden, 307

Handbook of Industrial Ecology. Cheltenham: Edward Elgar Publishing, 79-90.

Holck-Steen N., Bergsdal H., Brattebø H. (2007): Systems Analysis of Floor Area, Material and Life Cycle Energy Flows for the Norwegian Dwelling Stock in the Period 1960-2050. Building Research & Information (submitted for publication).

Kennedy C. A. (2003): Estimating the urban metabolism of Canadian cities: Greater Toronto Area case study. Canadian Journal of Civil Engineering Vol. 30, Part 2.

Kohler N., Schwaiger B., Barth B., Koch M. (1999a): Mass Flow, Energy Flow and Costs of the German Building Stock. http://www.ifib.uni-karlsruhe.de/web

Lowe R.J. (2001): Reducing Carbon Emission from the Building Stock – A Review of technical Potential, Barriers to Change and Policy Instruments. Paper commissioned for IEA/OECD Workshop on the design of Sustainable Buildings Policies, Paris, June 2001.

Meinel G., Herold H., Hecht R. (2006): Automatische Ableitung siedlungsstruktureller Grundlagendaten auf Basis digitaler Bildverarbeitung, GIS und räumlicher Statistik, In: Strobl, Blaschke, Griesebner (eds.), Angewandte Geoinformatik 2006, Beiträge zum 18. AGIT-Symposium Salzburg, 423-429.

Michel U., Ehlers M., Bohmann G., Tomowski D. (2006): Advanced Fusion Techniques for Automated Detection of Settlement Areas, in: Proceedings ISPRS Mid-Term Symposium: „From Pixels to Processes“ at ITC, the Netherlands, 2006.

Moll S., Bringezu S, Schütz S. (2003): Resource Use in European Countries. European Topic Centre on Waste and Material Flows. Copenhagen, March 2003.

Moll S., Bringezu S., Schütz S. (2005): Resource Use in European Countries. An estimate of materials and waste streams in the Community, including imports and exports using the instrument of material flow analysis. Wuppertal Report No. 1. Wuppertal: Wuppertal Institut.

Moll S., Vrgoc M., Watson D., Femia A., Gravgard Pedersen O., Villanueva A. (2006): A comparative European Study on Environmental Pressures Arising from Consumption and Production Patterns Employing Environmental Input-Output Analysis. Toronto.

Müller D. B. (2006): Stock dynamics for forecasting material flows – Case study housing in the Netherlands. In: Ecological Economics Vol. 59, No. 1, 142-159.

OECD (2005): Towards the Sustainable Use of Building Stock: Final Synthesis Report.

Rovers R. (2004): Existing Buildings, a Hidden Resource, Ready for Mining. Paper presented at the OECD workshop on Sustainable Buildings in Tokyo, January 2004.

Schiller G. (2007): Urban Infrastructure – Challenges for Resource Efficiency in the Building Stock. BRI, June 2007.

Siedentop S., Schiller G., Gutsche J.-M., Koziol M., Walther J. (2006): Siedlungsentwicklung und Infrastrukturfolgekosten – Bilanzierung und Strategieentwicklung. Forschungsprojekt im Auftrag des Bundesamtes für Bauwesen und Raumordnung sowie des Bundesministeriums für Verkehr, Bau- und Wohnungswesen. Final report. (BBR-online-publication).

Thinh N. X., Vogel R. (2005): Modelling Urban Land Use Dynamics with GIS and Cellular.

URBANMET (2007): Urban Fabric towards Sustainable Urban Metabolism, Proposal for ENV 2007.2.1.5.1. Area: Urban development, Coordinator Clemens Deilmann, with main contributions from Helge Brattebø, Stefan Brigezu, Paul Brunner, Susanne Kytzia.

8 Shaping Cultural Landscapes through Regional Governance

Ludger Gailing, Heiderose Kilper

8.1 A New Concept and its Place in the European Debate

In 2006 the German Standing Conference of Ministers responsible for Spatial Planning (MKRO) decided on three new concepts. The third concept can be summarised as "Conservation of resources, shaping of cultural landscapes". In addition to the usual spatial planning tasks such as giving careful consideration to competing claims for space and protecting open spaces and other natural resources, the concept includes a new component – to actively shape cultural landscapes. This spatial planning task involves more than just preserving places of historical importance since all urban, semi-urban and rural areas are regarded as cultural landscapes, subject to regional development concepts and strategies[1]. Cultural landscapes are viewed as a "qualitative supplement to traditional spatial development policies" and as "areas with their own particular characteristics serving the advancement of identification of the inhabitants with the region in which they live"[2]. Thus shaping cultural landscapes has, for the first time, become a "national task"[3].

This is of strategic importance for planning at *Länder* and regional levels, as the new task emphasizes the function of regional planning as a development tool and provides it with additional legitimacy in view of the emergence of other integrated space-related policy fields (such as river basin management following the EU Water Framework Directive or integrated rural development in LEADER regions).

The new concept of shaping cultural landscapes at regional levels alters the way in which such landscapes are dealt with in Germany. Cultural landscapes were previously regarded solely as public goods requiring protection. The preservation of evolved cultural landscapes is, for example, one of the spatial planning principles[4] laid down in the Federal German Spatial Planning Act (ROG) in 1998. The MKRO has thus entered debates that have played an important role in Europe in recent

1 cf. BMVBS (2006: 24-25)
2 BMVBS (2006: 25)
3 Alltschekow, Eyink, Sinz (2006: 12)
4 "Historical and cultural relationships and regional affiliations shall be maintained; the characteristic features and the cultural and natural monuments of evolved cultural landscapes shall be preserved." (§ 2 (2) no.13 ROG).

years. Two international documents, the European Spatial Development Perspective (ESDP) of the European Union[5], adopted in 1999, and the European Landscape Convention (ELC) of the Council of Europe[6], adopted in 2000, have defined these debates. In response to the continuous processes of globalisation, cultural landscapes are being rediscovered in these political documents as regional potential and an object of regional development.

The ELC states in its preamble that landscape is a "key element of individual and social well-being" and an "important part of the quality of life for people everywhere: in urban areas and in the countryside, in degraded areas as well as in areas of high quality" [7]. In the ESDP, natural and cultural heritage has been characterised as "economic factors which are becoming increasingly important for regional development"[8]. Consequently cultural landscape can be understood as an endogenous potential which influences not only the living and working conditions of the people of a region, but also its economic competitiveness and attractiveness for tourism[9].

The ESDP includes socio-cultural aspects in the term "cultural landscape" because "cultural landscapes contribute through their originality to local and regional identity"[10]. The ELC highlights the innovative aspect of the social construction of landscape and states that "'Landscape' means an area, as perceived by people, whose character is the result of the action and interaction of natural and/or human factors"[11].

Regarding the potential of cultural landscapes for regional development, both the ESDP and the ELC focus on the importance of quality of life that "plays an increasingly important role in the location decisions of new companies"[12]. Landscape "constitutes a resource favourable to economic activity and (...) can contribute to job creation"[13]. The image of "a distinctive landscape can be used to promote the qualities of an area for attracting new industry, for tourism and for other types of economic investment"[14].

The policy options outlined by the ESDP aim at the "preservation and creative development of cultural landscapes with special historical, aesthetical and ecological importance, (the) enhancement of the value of cultural landscapes within the framework of integrated spatial development strategies, (the) improved co-

---

5 European Commission (1999) referred to as "ESDP" in the following.
6 Council of Europe (2000) referred to as "ELC" in the following; the ELC, operative from 2004, has not yet been signed by the German Federal government.
7 ELC, Preamble
8 ESDP, Art. 134
9 cf. Röhring and Gailing (2006)
10 ESDP, Art. 151
11 ELC, Art. 1a
12 ESDP, Art. 134
13 ELC, Preamble
14 ESDP, Art. 323

ordination of development measures which have an impact on landscapes (and the) creative restoration of landscapes which have suffered through human intervention, including recultivation measures"[15]. The ELC further includes innovative measures such as the increasing of "awareness among the civil society, private organisations, and public authorities of the value of landscapes, their role and changes to them"[16]. Another aspect is the establishment of "procedures for the participation of the general public, local and regional authorities, and other parties with an interest in the definition and implementation of the landscape policies"[17].

While landscape planners are the target readership of the ELC, and the Federal Ministry for the Environment would thus be responsible for ratifying it in Germany[18], the ESDP is explicitly aimed at spatial planners. In this context, the Territorial Agenda of the European Union, which was approved in 2007, is of relevance as it basically confirms the approach taken in the ESDP: "The irreplaceable values of European ecological structures and cultural and natural heritage, especially cultural landscapes and the quality of design and process on architecture as well as the built environment, should constitute, against the background of the respective regional circumstances and potentials, the foundation for environmentally and culturally-oriented development which offers development perspectives, whilst safeguarding diverse cultural identities, particularly in regions that are lagging behind or undergoing structural changes"[19].

The new German concept of shaping cultural landscapes at regional levels is thus embedded in the context of a European debate in which cultural landscapes are no longer regarded solely as public goods to be protected by sectoral planning programmes, but increasingly as development potential within an integrated and qualitative regional development. At the same time, the recent debate on cultural landscapes in the social sciences has contributed to this new understanding. The debate is increasingly distancing itself from an essentialist[20] and elitist understanding of cultural landscapes and now places greater emphasis on their social constitution.

Cultural landscapes are viewed in the debate as a by-product or residual product of market forces, sectoral policies and their different institutional regimes on the one hand[21], and, on the other, as social constructs and representations of reality[22]. The latter perspective emphasizes in particular the identity-defining effect of cultural

15 ESDP, Art. 155

16 ELC, Art. 6

17 ELC, Art. 5c

18 cf. Hönes (2007: 45)

19 Informal Ministerial Meeting on Urban Development and Territorial Cohesion (2007: Art. 25).

20 The essentialist understanding of cultural landscapes is based on the assumption that such landscapes have an unchanging inherent value. However, this is inconsistent with the way in which such landscapes develop, which is process-based.

21 cf. Sieferle (2003); Apolinarski, Gailing, Röhring (2006)

22 cf. Head (2000); Norton (2000); Winchester, Kong, Dunn (2003)

landscapes, their importance for human patterns of perception, and the way they can be instrumentalised as images and brands, e.g. for the tourist industry or regional marketing. The debate on cultural landscapes is therefore now focusing on regional symbols, cultural values, and the meanings of landscapes, as well as the manifold forms of social and political governance to a greater extent than in the past[23].

8.2 Regional Governance as a Configuration for Shaping Cultural Landscapes and their Implementation by Means of Regional Parks

The new German spatial planning concept has twin aims for cultural landscapes, i.e. protection and conservation on the one hand and qualitative development and active shaping through policies on the other. This also requires a process of rethinking and a change in political and social practices. New forms of planning and steering regional development have become established in regional policy in the shape of many different forms of regional governance – prompted in no small part by the principles of European regional policies. They have shown themselves to be a governance configuration regarded as suitable for coping with the complex task of shaping cultural landscapes somewhere between protection and development.

Along with Mayntz[24], our assumption is that the concept of governance does not involve a change of paradigm but that a new, altered perspective has been adopted instead. Governance research and government theory focus on different aspects of reality. While the latter is predominantly concerned with the political-administrative system and includes a clear-cut separation between the subject and the object of government, this distinction disappears in governance research. Jessop[25] ascribes the popularity of the governance concept to the failure of markets and hierarchies in the face of the increasing complexity of society. In his view, governance provides possible advantages over the two other coordination mechanisms due to its potentially higher capacity for learning and innovation in a changing environment.

In the debate on governance a distinction is drawn between a narrow and a broad understanding of the term. The two approaches can be outlined as follows.

In the narrow sense of the term, "governance" is the opposite of "government", which denotes sovereign-hierarchical actions by an administration. Governance can be understood as a form of leadership or management which responds to the failure of markets and the state and fills the resultant gaps[26]. State institutions open up to cooperation with actors, groups and/or organisations in society and permit their having influence without exerting any pressure to cooperate with them. The state thus relinquishes its power to assert itself as a hierarchical, higher-ranking

23 cf. Donadieu, Périgord (2007)
24 Mayntz (2005)
25 Jessop (2002: 43)
26 cf. Fürst (2004)

institution and no longer decides unilaterally. It seeks to reach a consensus with those affected by its decisions and directives. This understanding of change in the state has also led to regions being "upgraded" as a level on which public tasks are fulfilled, resulting in a new level of state action and leadership which forms the institutional context of regional governance.

In the broad sense of the term, "government" and "governance" are not opposites. Simplifying dichotomies such as "governance vs. government" or "formal vs. informal" do not, it is assumed, do justice to the various, and frequently hybrid, forms of governance that can be confirmed empirically[27]. Institutional economists take this even further by speaking of three governance alternatives[28]. From a political science perspective, not only would new forms of governing such as public-private forms of cooperation or collaboration between non-state actors be described as governance but also classical forms of action by nation states. In this sense, the term "governance" includes many different forms of coordinating actions encompassing a broad spectrum of actors.

In addition to this, different views of social change within society can be identified in governance research. State-centred perspectives deal, above all, with the failure of the state in the face of changing social realities and derive top-down adjustments and modifications of state actions as normative conclusions.[29] On the other hand, critics of traditional state actions and the dominance of market-driven actions favour bottom-up approaches with participatory forms of governance[30] that can be initiated by different social forces.

Yet another understanding of the term can be gained by observing the development of the governance concept. It was initially a concept used in analysis to achieve a better understanding of the changes in political reality brought about by the interaction of public and private actors and by the multilevel European system. Its normative bias in the sense of "good governance" did not develop until later. Mayntz states that our concept of governance today is not only a broad and neutral one but also one which can be understood in the narrower sense as being at least latently normatively slanted[31]. Botzem also proposes splitting the concept into connected analytical and normative components in order to systematise the approaches to governance, as the latter are based on analysis but also postulate a desirable target condition[32].

In the meantime, the concept of governance with the normative bias referred to above has become established in regional development policies and thus in policies for cultural landscapes. Regional governance relies on collaborative processes involving both public and private actors and complements the sovereign-hierarchical

---

27 Blatter (2005: 121)
28 Williamson (1994: 102)
29 cf. Fürst (2004)
30 Jessop (2002: 58)
31 Mayntz (2005: 16)
32 Botzem (2002: 20)

processes of state and regional planning which also have to fulfil regulatory tasks and promote cohesion. Tasks related to cultural landscapes include

- integrating sectoral courses of action,
- supplementing approaches to preserving historical and cultural potential and the potential of natural areas with moves towards the economic valorisation and sustainable development of that potential,
- cooperation between town and country across *Länder* and regional borders and
- forging links between regional actors and initiating regional forms of collaborative planning.

These require action by the state, which, in support of regional governance, has to focus on cooperation, networks, and the self-organisation of society to a greater extent than on hierarchical coordination.

Regional parks and other project-oriented and cooperative strategies for the qualitative development of regions are practical examples of how the governance configurations required for regional governance are being put to the test. Regional parks and other networks of recreational spaces are mentioned explicitly as strategies for action within the new concept of preserving resources and shaping cultural landscapes. Successful examples are the Emscher Landscape Park in the northern part of the Ruhr region, the Rhein-Main Regional Park in the Frankfurt/ Main area and the regional parks around Berlin. The new governance structures do not present themselves as a substitute for classical formal planning instruments, but supplement them in their efficacy through models of stakeholder participation or their project orientation.

The idea of regional parks is an innovation in the German system of planning which tends to be based on restrictions and formal plans. Regional parks are project-oriented instruments of landscape-oriented regional management in city regions. The diverse means aimed at the valorisation of cultural landscapes are supposed to achieve the protection of open spaces in city regions, which has hitherto not been adequately ensured [33]. These not only include measures to open up and link open spaces but also those aimed at shaping them and promoting sustainable forms of land use. Urban landscapes are becoming an essential regional development factor, as open spaces are seen and shaped as cultural-historical and ecologically relevant places, recreational spaces, the basis for the quality of the location of a city region, and as a potential for urban agriculture.

Regional parks fulfil relevant regional governance functions aimed at shaping cultural landscapes[34]:

- Regional parks are supported by forms of action that permit regional processes of interaction and cooperation.
- Regional parks are spaces for collaborative regional development within city regions and thus overcome administrative constraints.

33 cf. Gailing (2005)
34 cf. Gailing (2007)

- Regional park development enables an improved consideration of local identities or aesthetic values in formal planning processes.

In the sense of multifunctional landscape management, individual open space interests are integrated and the status of open space is strengthened in a holistic approach. Former "residual space" can thus acquire a lobby in formal planning processes.

8.3 Institutional Requirements and Restrictions

The institutional requirements for regional governance as a means of shaping cultural landscapes are outlined as follows. They are based on practical experience with regional parks as a way of shaping cultural landscapes and other cooperative strategies for the qualitative development of regions, and on research into the institutional dimensions of cultural landscapes as regional common goods[35]. We discuss the restrictions arising from the characteristics of cultural landscapes as a by-product of multiple policies and the lack of control over the relevant informal institutions in their particular spatial contexts.

In the social sciences, institutions are understood as being generally recognised rules which form the basis of reliable patterns for the behaviour of actors. Institutions include social norms, legal provisions and distribution systems, established procedures, and patterns of actions and relationships[36]. Formal and informal systems of rules are combined in forms of interaction governed by institutions. Laws and administrative procedural rules are considered as formal institutions and as such can be shaped politically, while informal institutions such as traditions, cultural values or patterns of behaviour are generally not directly influenced by politics[37].

Seen from this perspective, the development of cultural landscapes is affected by a multitude of sectoral rules which have different aims and which influence human behaviour. Due to the pronounced heterogeneity and the diverse functions of cultural landscapes, a separate, complex system of institutions for dealing with them cannot exist, unlike in the case of other regional common goods.

Instead, cultural landscapes are the by-product of different systems of institutions, the coherent or diverging aims and inherent structures of which affect the focus of actors' activities. Owing to the complex intertwinement of these formal and informal, relatively strong or weak institutions, problems of interplay[38] may arise as a result of which institutions may impede each other's effectiveness.

Problems of institutional interplay arise from the different functions of cultural landscapes and must be resolved by means of adequate governance configurations.

35 cf. Apolinarski, Gailing, Röhring (2006); Röhring, Gailing (2006); Rodewald, Knoepfel (2005)
36 cf. Mayntz, Scharpf (1995)
37 cf. North (1990)
38 cf. Young (2002)

The disparate interests affecting cultural landscapes such as nature conservation, preservation of historical monuments and buildings, regional planning, agriculture, tourism, or economic development are reflected in formal institutional arrangements with very limited sectoral responsibilities – forming bureaucratic landscapes, so to speak[39]. Governance arrangements with the aim of institutionalising opportunities for cooperation between the sectoral systems of institutions are essential. Actors in civil society, land users, and other economically motivated actors must also be included in regional development processes aimed at shaping cultural landscapes.

Forms of governance which ensure regional cooperation and permit regional management are decisive for resolving problems of interaction with regard to cultural landscapes. Heterarchic structures shaped by networks, collective actors and active individuals are certainly capable of ensuring greater stability than monolithic structures. Multiple, highly differentiated and overlapping interaction arenas reflect the complex realities of cultural landscapes.

Cultural landscapes are the result of the individual and collective activities of actors and also form part of people's perception and identity. Active, decentralised actors therefore need to be supported and brought together in order to mobilise their development potential. Forms of social self-organisation (such as local heritage societies, artists' initiatives, networks of museums, tourism associations and cooperations between farmers, conservationists and providers of regional services) become a potential resource in cultural landscape policies. A particular challenge for governance configurations aimed at shaping cultural landscapes at regional levels is to develop new models of politics based on participatory inquiry and citizen-expert collaboration in order to bridge the divide between expert knowledge and the local interests of stakeholders[40].

The way in which informal institutions in particular influence the actors is of considerable importance for the development of cultural landscapes and consciously influencing the way in which they change. The following forms of informal institutions are particularly relevant for cultural landscapes as a common good:

- sectoral standards of value, concepts and patterns of behaviour,
- culturally defined representations and images of landscapes and
- specific characteristics of regional identities.

It is debatable whether informal institutions – as opposed to formal ones – can be influenced politically or even with regard to planning and if so, to what extent that is possible. For instance, regional identity can be understood as an attempt to construct a collective identity on the basis of a combination of primordial (historical, territorial etc.) factors or symbols and political or geographical boundaries[41], although such an identity is bound to elude direct attempts at controlling and instrumentalising it. On the other hand, attempts are being made to invent tradition[42], for example,

39 cf. Thomas (2003)
40 cf. Fischer (2000)
41 cf. Eisenstadt (1991)
42 cf. Hobsbawm, Ranger (1992)

and to support the building of regional identities by the same methods as used for maintaining cultural landscapes or by institutionalising cooperation-oriented arrangements in landscape governance. The approach of setting up regional parks discussed above, as well as other project-oriented and cooperative strategies for the qualitative valorisation of regions are good examples of the process of strategically instrumentalising informal institutions. In such forms of governance, public branding, image production, and stimulating regional identification processes, in other words "dramaturgical action" or actions that influence people's identities[43], are becoming increasingly important, compared to traditional planning.

Governance configurations for shaping cultural landscapes at regional levels with the aim of lending concrete form to the new German spatial planning concept must take the specific features of cultural landscapes, which are regarded as regional common goods, into account. The multiple functions of cultural landscapes, the heterogeneous nature of the actors, the particular role played by decentralised actors, and the formative significance of informal institutions which are difficult to control are challenges which the promoters of a regional policy geared towards cultural landscapes must meet. Shaping cultural landscapes at regional levels is an important task in Europe as a whole. With its specific approach of integrated, qualitative regional development based on the interests and identities of local stakeholders it complements previous perspectives of formal spatial planning.

43 Blatter (2005: 128)

References

Alltschekow P., Eyink H., Sinz M. (2006): Bewahren und Entwickeln. Neue Leitbilder der Raumentwicklung in Deutschland. In: Stadt und Grün, No. 12, 8-13.

Apolinarski I., Gailing L., Röhring A. (2006): Kulturlandschaft als regionales Gemeinschaftsgut. Vom Kulturlandschaftsdilemma zum Kulturlandschaftsmanagement. In: Matthiesen U., Danielzyk R., Heiland S., Tzschaschel S. (eds.), Kulturlandschaften als Herausforderung für die Raumplanung. Verständnisse – Erfahrungen – Perspektiven, Forschungs- und Sitzungsberichte der ARL, No. 228. Hannover, 81-98.

Blatter J. (2005): Metropolitan Governance in Deutschland: Normative, utilitaristische, kommunikative und dramaturgische Steuerungsansätze. In: Swiss Political Science Review, Vol. 11, No. 1, 119-155.

BMVBS (2006): Leitbilder und Handlungsstrategien für die Raumentwicklung in Deutschland. Berlin: Geschäftsstelle der Ministerkonferenz für Raumordnung.

Botzem S. (2002): Governance-Ansätze in der Steuerungsdiskussion. Steuerung und Selbstregulierung unter den Bedingungen fortschreitender Internationalisierung, discussion paper FS I 02-106. Berlin: Wissenschaftszentrum Berlin für Sozialforschung.

Council of Europe (2000): European Landscape Convention (ELC). Firenze.

Donadieu P., Périgord M. (2007): Le Paysage entre Natures et Cultures. Paris: Armand Colin.

Eisenstadt, S. M. (1991): Die Konstruktion nationaler Identitäten in vergleichender Perspektive. In: Giesen B. (ed.), Nationale und kulturelle Identität. Studien zur Entwicklung des kollektiven Bewusstseins in der Neuzeit. Frankfurt am Main: Suhrkamp, 21-38.

European Commission (1999): European Spatial Development Perspective (ESDP). Luxemburg.

Fischer F. (2000): Citizens, Experts, and the Environment. The Politics of Local Knowledge. Durham, London: Duke University Press.

Fürst D. (2004): Regional Governance. In: Benz A. (ed), Governance – Regieren in komplexen Regelsystemen. Eine Einführung. Wiesbaden: VS-Verlag, 45-64

Gailing L. (2005): Regionalparks – Grundlagen und Instrumente der Freiraumpolitik in Verdichtungsräumen. Dortmunder Beiträge zur Raumplanung, No. 121. Dortmund.

Gailing L. (2007): Landscape governance in urban agglomerations – The case of regional parks in Germany. In: Kenneweg H., Tröger U. (eds.), 2nd International Congress on Environmental Planning and Management. Visions – Implementations – Results, Landschaftsentwicklung und Umweltplanung, Schriftenreihe der Fakultät Planen Bauen Umwelt, No. S20. Berlin, 481-484.

Head, L. (2000): Cultural Landscapes and Environmental Change. London: Hodder Arnold.

Hobsbawm E., Ranger T. (eds., 1992): The Invention of Tradition. Cambridge: Cambridge University Press.
Hönes E. R. (2007): Rechtliche Aspekte zum europäischen und deutschen Kulturlandschaftsschutz. In: Bauerochse A., Haßmann H., Ickerodt U. (eds.), Kulturlandschaft. administrativ – digital – touristisch. Berlin: Erich Schmidt Verlag. 31-90.
Informal Ministerial Meeting on Urban Development and Territorial Cohesion (2007): Territorial Agenda of the European Union. Towards a More Competitive and Sustainable Europe of Diverse Regions. Leipzig.
Jessop B. (2002): Governance and Meta-governance in the Face of Complexity: On the Roles of Requisite Variety, Reflexive Observation, and Romantic Irony in Participatory Governance. In: Heinelt H., Getimis P., Kafkalas G., Smith R., Swyngedouw E. (eds.), Participatory Governance in Multi-Level Context: Concepts and Experience. Opladen: Leske & Budrich, 33-58.
Mayntz R. (2005): Governance-Theory als fortentwickelte Steuerungstheorie? In: Schuppert G. F. (ed.), Governance-Forschung: Vergewisserung über Stand und Entwicklungslinien. Baden-Baden: Nomos, 11-20.
Mayntz R., Scharpf F. W. (1995): Der Ansatz des akteurzentrierten Institutionalismus. In: Mayntz R., Scharpf F. W. (eds.), Gesellschaftliche Selbstregelung und politische Steuerung. Frankfurt am Main, New York: Campus, 39-70.
North D. C. (1990): Institutions, Institutional Change and Economic Performance, Cambridge, New York, Melbourne, Madrid: Cambridge University Press.
Norton W. (2000): Cultural Geography: Themes, Concepts, Analyses. Oxford: Oxford University Press.
Rodewald R., Knoepfel P. (2005): Institutionelle Regime für nachhaltige Landschaftsnutzung. Régimes institutionels pour le développement durable du paysage, Oekologie und Gesellschaft, No. 20. Zürich, Chur: Verlag Rüegger.
Röhring A., Gailing, L. (2006): Cultural landscapes as a potential force for regional development – theoretical approaches and case studies in East German suburban and rural regions. In: Komornicki T., Czapiewski KŁń. (eds.), Regional Periphery in Central and Eastern Europe, EUROPA XXI, 15. Warsaw: Institute of Geography and Spatial Organization, Polish Academy of Sciences, Polish Geographical Society, 13-28.
Sieferle R. P. (2003): Die totale Landschaft. In: Oswald F., Schüller N. (eds.), Neue Urbanität – das Verschmelzen von Stadt und Landschaft. Zürich: gta, 59-76.
Thomas C. W. (2003): Bureaucratic Landscapes. Interagency Cooperation and the Preservation of Biodiversity. Cambridge, London: The MIT Press.
Williamson O. E. (1994): Transaction Cost Economics and Organization Theory. In: Smelser N., Swedberg R. (eds.), The Handbook of Economic Sociology. Princeton, 77-107.
Winchester H., Kong L., Dunn K. (2003): Landscapes – Ways of imagining the world. Harlow: Pearson.
Young O. (2002): The Institutional Dimensions of Environmental Change. Fit, Interplay, Scale. Cambridge, London: The MIT Press.

9 The German-Polish Border Region from a German Perspective – quo vadis?

Sebastian Lentz, Günter Herfert, Annedore Bergfeld

Introduction

Cross-border relations between Germany and Poland should not be considered without taking into account the special history of their mutual border, even in 2007. After many years of violent territorial shifts, armed conflicts, expropriations, forced migrations, expulsions, and relocations on both sides, and after a long period of political uncertainty in the aftermath of all this, the constitutional basis for the development of regional co-operation across the national border was not laid until the German-Polish border treaty of 1990, after the disintegration of the confederation of socialist states in Eastern Europe. Such co-operation forms one of the most important regional political instruments for strengthening inner cohesion in the European Union, intended to facilitate a neighbourly coexistence and improve the chances of interaction of the population on both sides of the border. It is not least a matter of a relationship of trust in the immediate border regions, which, in this case, should have a stabilising effect on the political relations of the neighbouring states. At the same time, the social and economic development of the often structurally weak border regions should be stimulated by, for example, complementary infrastructure becoming useful for all the inhabitants of the border region. Poland's entry to the European Union in 2004 again improved the preconditions for such co-operation, as the instruments of EU regional structural politics can now become effective.

Co-operation in border regions between Germany and Poland takes place on a variety of levels. The most important is co-operation in civil society, meaning thc direct contacts of citizens in associations, initiatives, or in spontaneous encounters. Co-operation has developed between communities and at the county level in a positive manner and continues to proceed in the region close to the border, in particular under the umbrella of the Euroregions. Furthermore, there is an active regional co-operation between the German federal states and the neighbouring Polish *voivodships*. The national governments of Germany and Poland in turn define the legal framework conditions for co-operation in border regions to a great extent. In practice, institutional co-operation is still impeded repeatedly by the widely differing powers of the agents on the ground resulting from the federal structure of Germany and the centralistic organisation of Poland.

The German-Polish border is, historically, a special place, compared to the other outer borders of Germany: It was only finally recognised as the western border of Poland by unified Germany in the German-Polish border confirmation treaty of 11th November 1990. The resulting debates in both Germany and Poland showed that this was, at the time, more than just a formal act. The Polish publicist Adam Krzemiński described the special case of the German-Polish border as: "A completely artificial border. It divides regions which were joined together for centuries". The area around the Oder and the Neisse has only been a *border* region since 1945; people on both sides lacked experiences of living at a border. Precisely this is the essential difference from border regions with a longer experience and a historically grown acceptance of both separation and togetherness, such as is the case on the German-Belgian border. This is also complicated by the fact that not only the outer border of two states was moved, but people also were "displaced". For example, the area West of the Oder is now inhabited by many displaced people from the former Neumark, which, as "Ziemia Lubuska" (Lubusz Land), was completely reoccupied by displaced Poles from the East and new settlers from central Poland (Mildenberger 2007: 2ff.).

Three years after Poland's entry into the European Union, a positive balance can be recognized from the perspective of German-Polish co-operation in the border regions. German fears in anticipation of EU expansion, which were often subject to populist stoking, have not been confirmed: Cross-border criminality has not increased, pressure on the employment market has not intensified, especially not in the structurally weak eastern German border region. On the contrary, complaints are increasingly being heard from certain sectors seeking qualified employees from Poland. The skilled craftsman sector already co-operates successfully across the border. Competition has only increased in the services sector, but does not by any means occur in just the one direction. Nor have Polish fears in connection with entry to the European Union come true. Rather, many Polish farmers now profit from membership of the EU. Approval rates in the country towards the EU have, in the meantime, reached record highs – much more than before entry to the EU (Mildenberger 2007: 5).

Fig. 1: The German-Polish border region, 2007.

In spite of this, the German-Polish border region (fig. 1) will probably remain a structurally weak region for a long time, with low economic power and aggravated demographic problems on both sides of the border. Many forecasts on the development of the border region, which for its part is characterised by highly pronounced disparities both jointly between the German and Polish border regions and separately within the German or Polish part of the border region, have proved over-optimistic. It was thought that the advantages of the new border position, the differences in price of goods and services, and the differences in labour, raw material, and energy costs would attract foreign investors (Kaczmarek and Stryjakiewicz 2006). This area is usually however "leapfrogged" by cross-border investors and can only profit a little from economic contacts between West Germany and Berlin on the one hand and Polish metropolitan regions on the other (Krätke and Borst 2004). In this context, increased importance is attached to the development of endogenous potential in direct, cross-border co-operation. Positive approaches can be demonstrated and expanded in particular in the regions close to the border.

The present article uses select core themes (demography, employment market, transport, and education) to primarily address the existing regional disparities of the German part of the German-Polish border region, but with an eye on the entire border region. It concludes with an outlook on possible developmental trends.

9.1 Co-operation and Links in the Border Region

In the border region, numerous cross-border co-operation and functional links already exist which can be used as a basis for further and improved co-operation between the two nations in particular and specifically for the regions in immediate proximity to the border.

A considerable proportion of cross-border co-operation is formed by the four Euroregions (fig. 1), which are active in the fields of business development, transport, education and training, and cultural exchanges, among others. In the immediate border regions, they are important decision-makers for the coordination of planning and projects affecting the border region. Spatial development on both sides of the border can be positively influenced by coordinated and jointly developed projects. An expectation in the Euroregions is also that they contribute towards overcoming cultural, mental, and linguistic barriers in the population of the border region and thus potentially encourage the formation of awareness on both sides of the chances of solving problems together.

However, co-operation is not without impediments, although both sides could benefit more from each other (Kaczmarek and Stryjakiewicz 2006). Co-operation is insufficient between the Euroregions, the formal planning levels, and other important regional agents. The difficulties are partly due to differing thematic priorities on both sides of the border, owing to dissimilar problem contexts and spheres of influence of agents. Existing working groups on the various planning levels have contributed to an improved coordination of ongoing projects but must accept how priorities are set on political levels, so that the result is often an insufficiently integrative treatment of the region. Correspondingly, the Euroregions encounter barriers in the perception of their function as developers of the whole region and project initiators

(Leibniz Institute for Regional Geography et al. 2006). An indication of this is that many of the projects having received financing until 2006 in the European development programmes PHARE or INTERREG were hardly cross-border or were often implemented parallel to each other (Kaczmarek and Stryjakiewicz 2006). A new quality of project work should however be achieved within the development period from 2007 to 2013 (Objective 3 development) and the setting up of mutual Technical Secretariats on the German or Polish side. The implementation targets of the Objective 3 programme place high demands on project partners with regard to the coordination and planning process, so that previous, often local or small regional project organisers such as associations may very well reach the limits of their capabilities and possibilities.

Around 10,000 German companies currently maintain long-term business relations in Poland. Strong sectors are the automotive industry, financial services, wholesale and retail, food processing, and energy supply. At the same time, the number of Polish companies becoming established in Germany is increasing. Approximately 20,000 Polish companies have been registered in Germany since Poland's entry to the EU on 1st May 2004 (DPERON 2007: 20). However, the economic networkings of Polish economic centres and large corporations are ostensibly conducted with companies in the former West Germany and Berlin. This is also demonstrated by the fact that 94% of the German companies active in Poland have their office headquarters in western Germany (Krätke and Borst 2004).

In the immediate border region, up to 2004, the economic contacts, acting as driving forces for integration processes in many other border regions, were adversely affected by the outer border of the EU. This is due to the lack of export orientation of companies on the German side, as well as by the dominant agricultural sector in large areas of the Polish border region and the associated low potential for development (Hamburg Institute of International Economics 2006). Since integration into a common market, interest in cross-border investments has increased on both sides. By locating in the German border region, Polish investors see the possibility of tapping into state development options and the employment market, but also the opportunity of being able to provide their products with the "Made in Germany" image. Moreover, the motives of expanding supply and service relationships also have a positive effect. The promotion of contacts between German and Polish companies in the border region is, in particular, borne by the German-Polish Economic Development Corporation in Gorzów Wielkopolski and the regional offices of the Chamber of Industry and Commerce. In addition, advisory bureaus of the German Chambers of Industry and Commerce (DIHK) were opened in Görlitz and Szczecin.

On a local level, the twin cities of Frankfurt(Oder)/Słubice, Guben/Gubin and Görlitz/Zgorzelec, which each formed a single unit before the Second World War, and the "Little Triangle" city co-operative of Zittau/Bogatynia/Hradeck nad Nisou play a special role, since the interest in co-operation is potentially at its highest here, despite economic disparities and differences in the mentalities of the population.

The co-operation between Frankfurt and Słubice in particular has, after 1990, been considered a model for the renewal of German-Polish co-operation (Kaczmarek and Stryjakiewicz 2006). In 1991 an agreement was made on social and economic co-operation, and in 1994 a common "spatial structural concept" was created. In addition, joint marketing measures were carried out.

Frankfurt (Oder)
Polish-customer-oriented retail traders and services in 2003

Kind of facilities	Number of facilities	among them with			
		Labels in Polish language	Prices in PLN	Payment in PLN	Service in Polish language
Retail trade	20	9	–	13	1
Services	6	4	1	2	2
Total	26	13	1	15	3

Słubice
German-customer-oriented retail traders and services in 2003

Kind of facilities	Number of facilities	among them with			
		Labels in German language	Prices in €	Payment in €	Service in German language
Retail trade	55	40	15	46	53
Services	33	29	8	27	31
Total	88	69	23	73	84

Tab. 1: Commercial and service operations in Słubice and Frankfurt/Oder oriented towards German and Polish customers. Source: Kaczmarek and Stryjakiewicz 2006.

With Poland's entry to the EU, an upturn in the service industry has taken place on the Polish side. Whereas many commercial operations on the Polish side are oriented towards German customers, up to 2003, an orientation towards Polish customers as in Frankfurt was still the exception to the rule (fig. 2). In general the contacts between the two cities have a predominantly formal character, however the quality of co-operation as well as enthusiasm for working towards integration into a "European city" should be considered separately. Obstacles to intensification are, in addition to the political background, negative stereotypes prevalent in both societies, but also the very different economic potential of the cities.

Various small steps towards rapprochement on different levels (sports, culture, education, business) gradually create greater understanding but will not eliminate many problems even in the medium term. Studies among youths in the twin city of Guben-Gubin have for example shown "that the optimistic opinion that the young generation would reinterpret the border for itself beyond historical burdens and overcome obstacles to interaction and communication in the border region is justified only to a limited extent" (Hoorn 2006: 30). Although there is potential for cross-border understanding, active political support and promotion, in particular of the further expansion of various local initiatives (e.g. German-Polish nurseries), is necessary for a real rapprochement. Correspondingly, the German-Polish Congress of Local Self-governing Bodies [Deutsch-Polnische Kongress Lokaler Selbstverwaltungen] in Piechowice in November 2007 (www.kas.de/proj/home/pub/48/1/year-2007/dokument_id-12673/index.html) called for the

- activation of committees within the framework of the work of the German-Polish Government Commission for regional and border co-operation,
- initiation of new projects for education and social communication as a basis for mutual understanding,
- increased financial support for youth exchanges and a system for promoting Polish language courses for Germans.

The obligation introduced in 2007 to have international relations at the voivodship, county, city, and community level approved by the Foreign Office has led to the fact that coordination processes have become more time-consuming. This political framework condition has blocked or impeded many projects particularly in the border regions (DPERON 2007: 18). In addition, it should not be forgotten when assessing cross-border co-operation, that often regional or local interests are in the focus of attention, and cross-border harmonisation is not always considered a priority and absolutely necessary.

9.2 Select Structural Aspects in the Border Region

Demography

The regions on both sides of the border have the tendency towards growing regional polarisation in demographic development in common. The contrast between a small number of regions experiencing growth and extensively declining regions is becoming sharper - and this trend will intensify further according to existing forecasts (Hamburg Institute of International Economics 2006). The two border regions will differ in one respect, however:

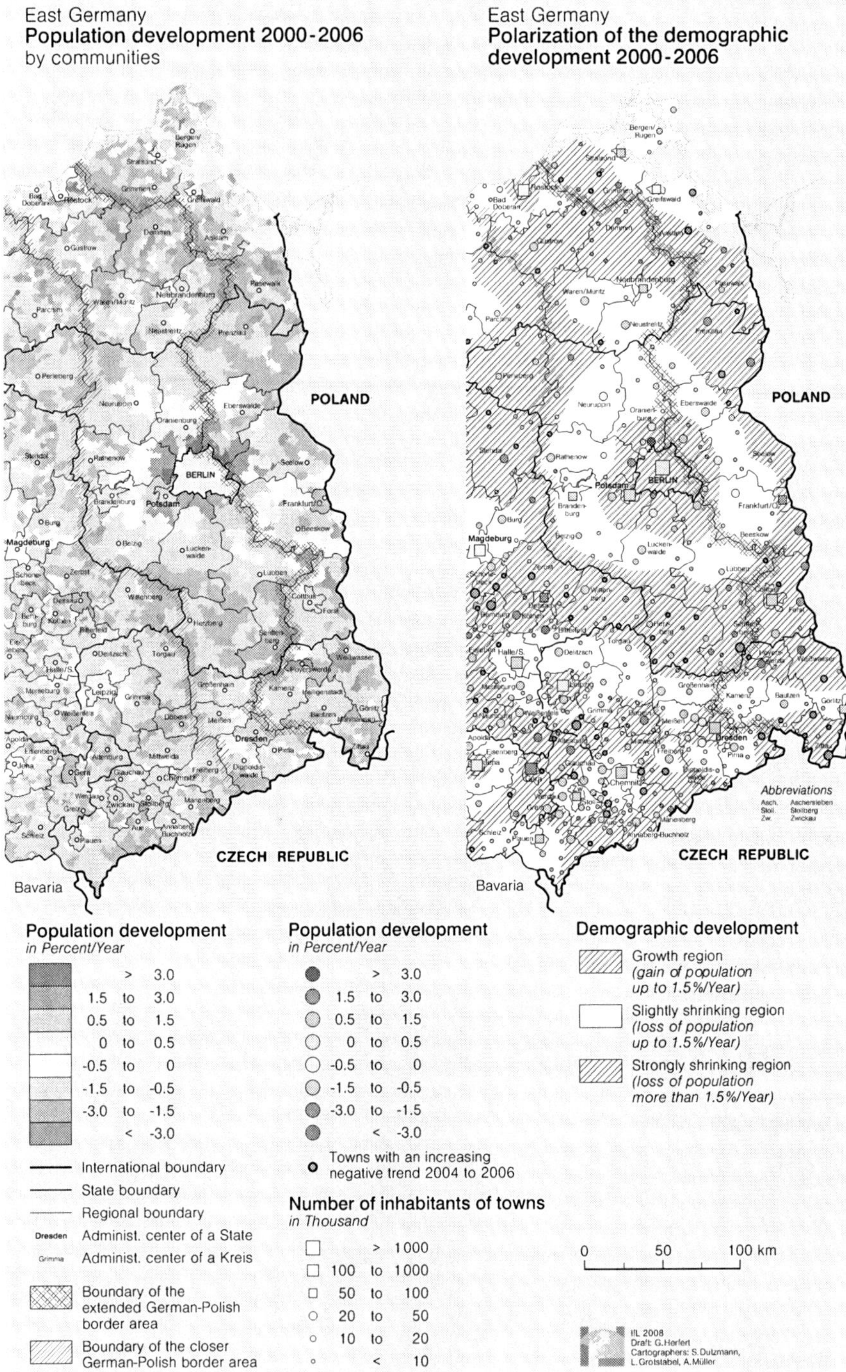

Fig. 2: Population development and regional polarisation of demographic development in the German part of the German-Polish border region, 2000-2006. Source: Herfert 2007, Regional Statistical Offices.

Whereas suburbanisation will continue on the Polish side surrounding the cities of Szczecin, Poznan and Wroclaw, the dynamic suburbanisation wave of the 1990s has largely petered out in eastern Germany, except in the region surrounding Berlin (Herfert and Lentz 2006). A fundamental change in spatial development has occurred here: With Berlin, Dresden and Rostock – and also very recently Greifswald – some large centres in the German border region again report slight population gains since 2000 after a long phase of decline (fig. 3). In these centres, where business and the knowledge economy are connected to a positive image of the city, a slight trend towards reurbanisation is apparent. It is based less on remigration from the surroundings, but rather primarily on migration from other regions of eastern Germany. Supporting this development are in particular young people migrating for reasons concerning education and looking for their first job, who considerably alleviate the structurally dictated ageing processes in the cities. Generally speaking, it can be assumed, owing to the growing attractiveness of these centres, that they will continue to be the target of interregional migration flows, also of more highly qualified workers. This may stabilise their "reurbanisation path" further. After more than a decade of great population losses to West Germany, remigrations from the former Federal Republic could also be conceivable in the future. However, under the present framework conditions, reurbanisation will remain an isolated phenomenon of a few centres in eastern Germany.

In contrast to the islands of demographic growth, the rest of the border region is affected by severe demographic decline. The regions close to the border in particular are some of those worst affected by out-migration in eastern Germany. A negative demographic trend has become established here, a downward spiral of lack of growth, job losses, decline in demand, ageing population, and out-migration, which might intensify further (Herfert 2007).

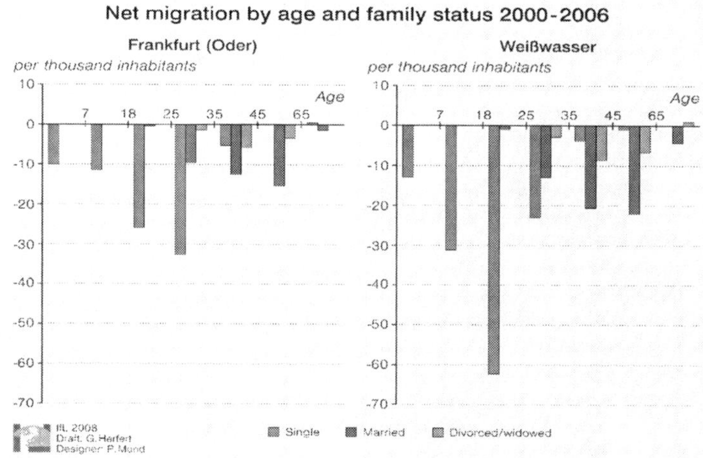

Fig. 3: Net migration (in ‰) by marital status in Frankfurt (Oder) and Weisswasser, 2000-2006. Source: Herfert 2007, Regional Statistical Offices.

According to the 2020/2050 spatial planning prognosis (Federal Office for Building and Regional Planning 2006), a double shift in age structure requires observation: A decline in school age children combined with a great increase in the elderly. Particularly affected are small and medium-sized towns, numerous former county and industrial towns such as Weisswasser, which lost their economic base after reunification (fig. 4). Even large towns with universities and technical colleges such as Cottbus, Frankfurt/Oder and Neubrandenburg are subject to this downward spiral. They are suffering losses in particular of young people, who migrate towards western Germany or to the eastern German growth centres after completing their education. An alleviation of this negative development through influxes from Poland is currently only marginal, except for the migration gains of some small districts of the county of Uecker-Randow in Mecklenburg-Vorpommern beginning in 2005 (tab. 4). These are districts in the immediate border region and comprise a good transport situation with regard to Szczecin, as a result of which they are becoming more attractive as a place of residence for Polish citizens.

In general, and this also applies to the twin cities or those close to the border, the expansion of the EU to the East however has hardly had any effects on the housing markets in the eastern German border region to date (Leibniz Institute for Regional Geography et al. 2006). With Poland's entry to the Schengen Agreement, and the thus altered framework conditions, incipient developments will be closely watched. With approximately 326 housing units per 1,000 inhabitants, Poland has the lowest quota of the EU in this regard. This will not change fundamentally in the coming years, even with maintaining current housing policies. Considering the current situation (tab. 2), 1/5 of all persons asked in a survey in Bogatynia (2005) could envisage moving to the neighbouring German town of Zittau, provided the rents were affordable. The interest of young people with a high level of education was particularly high. Facilitation of border transport coupled with a simultaneous equalisation of wages could therefore lead to a higher demand for housing in the German regions close to the border.

Zittau (D) and Bogatynia (PL)
Dwelling stock 2004

	Zittau	Bogatynia
Number of apartments	17610	8338
Vacancy	4388 *(25%)*	ca. 150 *(2%)*
Residents per apartment	1.6	3.1
Living space in square meters per apartment	65	58
Living space in square meters per capita	38	19

Tab. 2: Comparison of housing stock in Bogatynia and Zittau, 2004. Source: Municipal authorities of Bogatynia and Zittau.

The trends towards demographic polarisation define all areas of public life in the border region and require completely new strategies for action. For the severely declining regions the question of ensuring equivalent living conditions arises, and it is still imperative "to guarantee equal access to public facilities and services with minimum quality standards and within a reasonable distance for the populations of all regions in the future as well" (Sinz 2006: 610). Even if these minimum quality standards are still worded very vaguely in the new principles for spatial development in Germany (Federal Ministry of Transport, Building and Urban Affairs 2006), it might generally be difficult to maintain them to their present extent in the severely declining regions. This is due to the fact that the population of these regions would decline by more than half its current rate by 2050 if the increased exodus of young, mobile people continues. As a strategy to counteract the above-mentioned "downward spiral", the new principles suggest the increased utilisation of endogenous potential as well as the formation of solid partnerships between growing and declining areas. A stabilisation based on particular regional strengths (e.g. natural attractions, economic potential) or on a reorientation towards regionally directed economic cycles might be possible in some regions.

Employment Market

The sub-regions are currently faced with comparatively great economic problems. The structural weakness on the German and Polish side and the insufficient funds of the majority of the communities along the border impede long-term economic development.

All regions of the area under consideration were able to generate a rising gross domestic product in recent years - but at different levels. Whereas the German part of the area under research has a GDP of 16,300 to 21,700 Euro per capita, which corresponds to approx. 75% of western German levels, there is an even lower GDP of around 9,700 to 11,100 Euro per capita in the Polish part.

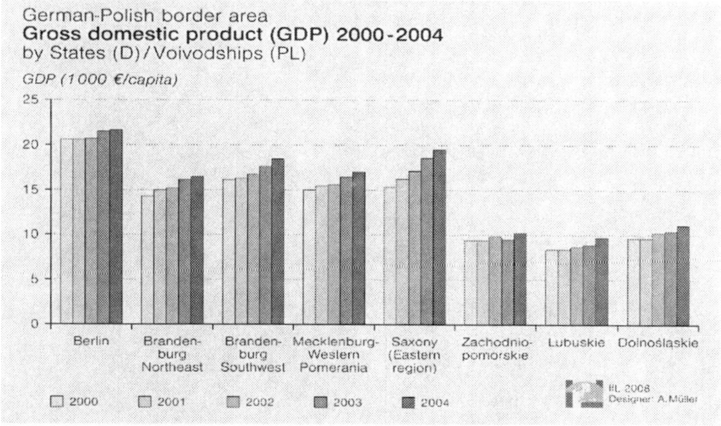

Fig. 4: Growth of GDP in the German-Polish border region 2000-2004. Source: DPERON 2007: 25.

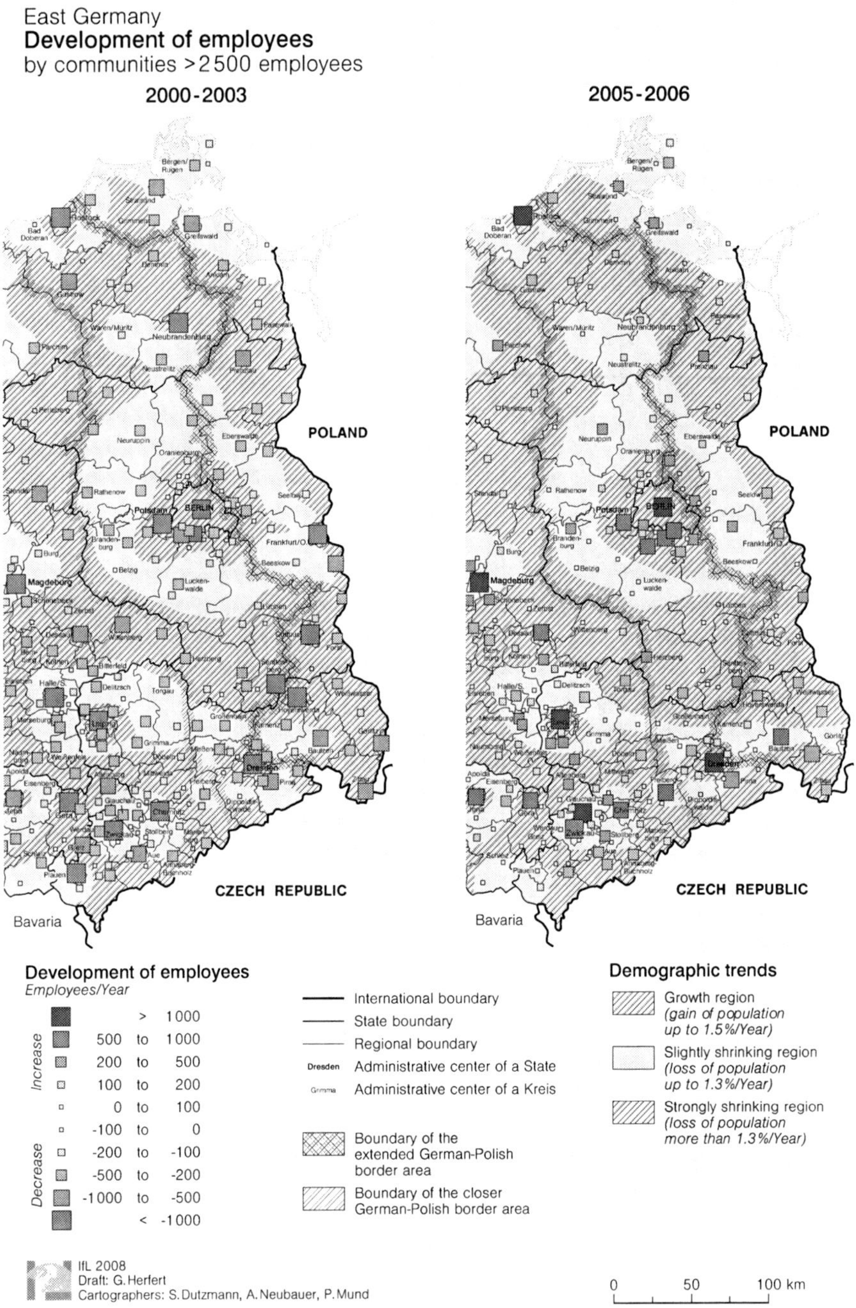

Fig. 5: Growth of employees paying social insurance contributions in the German part of the German-Polish border region 2000-2003 and 2005-2006. Source: Federal Job Agency, Nuremberg.

In all, the border region is characterised by an insufficient business density, a weak export basis, a lack of capital and technology, and direct investment which flows almost exclusively into the urban centres of Berlin, Posznan, and Wroclaw but omits the region close to the border.

Correspondingly, the situation on the employment market on both sides of the Oder and Neisse is characterised by high unemployment, with the regional differences in Poland (10-40%) being even more serious (Hamburg Institute of International Economics 2006). Whereas in rural areas almost every fourth person is unemployed, the large centres and their surrounding regions distinguish themselves by lower unemployment rates: on the Polish side Szczecin, Posznan, and Wroclaw, on the German side Dresden, Berlin, and Rostock. Not even the economic upturn which began in Germany in 2006 has changed the regional disparities in the German border region. While the three cities profit from the upturn and have achieved growth in employment for the first time since unification, no relief can yet be seen on the employment market in the rest of the border region; job losses have only been reduced or stopped (fig. 6).

In addition, the employment rate in the Polish part of the border region is much lower than that of the German part. A fundamental change in this employment situation is not to be expected in the future, owing to the existing framework conditions. These regions are in danger of being detached from the general upward trend in Germany as a result of the constant out-migration of qualified workers. Particularly problematic for the future employment market might be the increasing out-migration of mobile and well-educated young people owing to the lack of opportunities for labour market entry. To what extent demographic decline, in particular the erosion of the cohorts after 1990, will lead to an easing of the first employment market in the border region is debatable, since it is primarily a case of structural unemployment.

A significant proportion of employment market links in the German-Polish border region are currently still made within the framework of illegal or semi-legal employment conditions in the form of underground and bazaar economies (Leibniz Institute for Regional Geography et al. 2006). The official commuting links into the Polish border region are only marginal (tab. 3), on the one hand because of the regressive economic development in the German region close to the border, on the other hand because of the "2+3+2" model still used by Germany to restrict employee freedom of movement[1]. This limits further networkings in the employment markets.

On 1st November 2007, the federal government opened up the employment market to engineers from the EU new member states in mid- and eastern Europe (Becker 2007:

1 In all, the existing legal provisions offer different options for Poles to work in Germany. Whereas most work permits are however granted only in defined areas and with time limits, border commuter status offers the option of working in Germany, legally and under the same conditions according to employment law as German employees, to people who do not fall into these job groups as well as to unqualified laborers.

Selected German cities in the nearer border area
Commuters from Poland 2000 and 2006

City	In-Commuters 2000			In-Commuters 2006		
	total	male	female	total	male	female
Pasewalk	0	0	0	10	4	6
Schwedt (Oder)	0	0	0	73	69	4
Bad Freienwalde	0	0	0	10	5	5
Frankfurt (Oder)	50	34	16	47	32	15
Görlitz	37	27	10	38	24	14
Zittau	20	15	5	10	7	3

Tab. 3: Commuters from Poland to Germany in the region close to the border, 2000 and 2006. Source: FJA, Nuremberg.

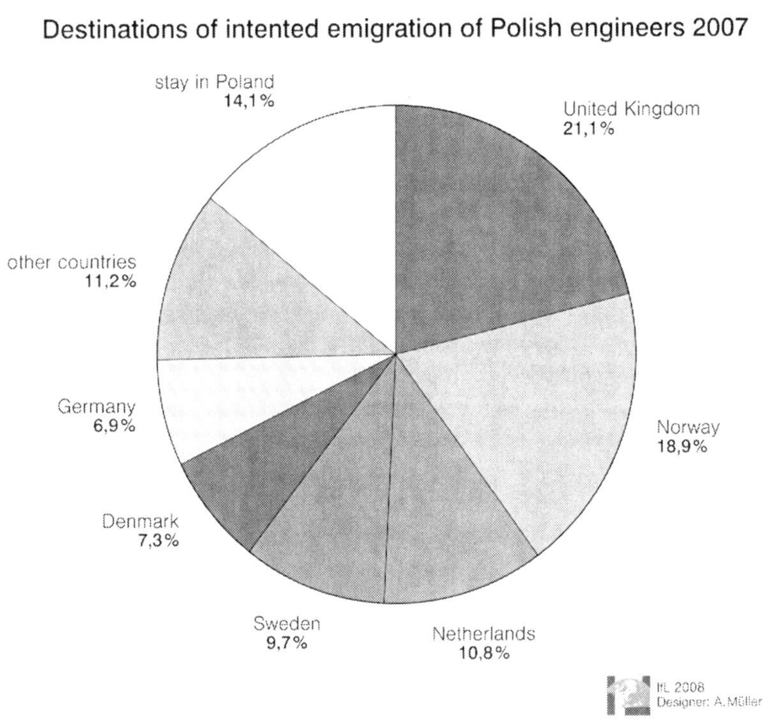

Fig. 6: Emigration destinations of Polish engineers (in %). Source: Bank Danych o Inżynierach – BDI (Databank of Engineers) (www.bdi.com.pl).

2). Germany was thus reacting to the lack of skilled workers at the time. In practice, however, the effect on Polish migrants will remain quite low. The United Kingdom and Ireland are much more attractive for skilled Polish workers. In a survey carried out by the BDI approximately 14 days after the decision, Germany is actually the country to which skilled Polish workers are least attracted. Of the 1,150 engineers surveyed by the company, only 6.9 percent wanted to move to their western neighbour (fig. 7).

A further important reason why there will hardly be a mass exodus to the West is the existing shortage of skilled labour in Poland (Wóycicka 2007: 2ff.). New links can be seen since 2004, but only on a low level. For example, apprenticeships of Poles in Germany increased. This can be observed without exception in the border region as well.

If one considers economic migration between Poland and Germany as a whole (tab. 3), it can be seen that the region close to the border is largely skipped. Economic migration is predominantly directed towards Berlin and western German growth regions. A fundamental change in the migration pattern for Polish migrants is only to be expected in Germany after full freedom of movement is granted (Glorius 2007).

German towns and regions (Kreise) in the German-Polish border area
Migration balance with Poland 2000-2006

	2000	2001	2002	2003	2004	2005	2006
Berlin	1364	1209	1789	2128	4437	6189	4293
Kreis Uecker-Randow	22	16	16	17	28	142	228
Dresden	-163	56	128	85	221	263	225
Rostock	32	1	76	18	-49	14	200
Görlitz	45	79	132	85	122	79	94
Frankfurt (Oder)	182	162	192	42	-91	-76	90
Schwedt (Oder)	16	22	37	49	27	-26	18
Guben	18	17	18	-10	-12	11	12
Cottbus	85	39	76	10	84	20	-52
Germany	**22696**	**24501**	**22229**	**22014**	**34745**	**53666**	**51151**

Tab. 4: Net migration (persons) of cities/towns and counties in the German part of the German-Polish border region with Poles, 2000-2006. Source: Regional Statistical Offices.

For the further development of employment market links, further cross-border inhibitions such as the lack of recognition of professional qualifications, language barriers, differing tax and legal systems and market entry costs should however be taken into account (Leibniz Institute for Regional Geography et al. 2006: 45).

The region close to the border is currently - except for the Szczecin region - an area of depletion, both from the German and the Polish point of view. Economic stabilisation of this region cannot be achieved solely on the basis of endogenous potential. Against the background of a declining and ageing population and a trend towards knowledge-based economic structures, development priorities should therefore be concentrated on the urban regions, in which the peripheral regions could then participate (Hamburg Institute of International Economics 2006). Such beacon effects could radiate from Berlin, Dresden and Rostock on the German side and from Szczecin on the Polish side towards the region close to the border.

An essential element for the stabilisation and development of the region close to the border would be a common economic and employment market, for which the dismantling of employment market barriers is however a prerequisite. Within the border region, there should be an orientation towards regional core competences and core development themes, with twin cities receiving particular consideration from a spatial as well as strategic perspective. Despite any competition, cross-border regional marketing, simplified and harmonised permit and planning procedures, and a common guarantee of quality and safety standards should be worked towards by all means. With regard to the expected skills shortage, the setting up of bilingual, central employment market and business development institutions and mutual recognition of qualifications will be necessary to prevent further brain drain (Leibniz Institute for Regional Geography et al. 2006).

Taking these circumstances into account, the objective of achieving closer economic and infrastructural co-operation of the Oder border regions by means of agreements and projects, but without new administrative structures, was pursued with the founding of the Oder Partnership in 2006. This partnership is understood as an information and discussion platform featuring innovation and technology transfer, improvement of cross-border transport connections, and tourism as essential spheres of activity. In the same manner, it is hoped that many individual initiatives, e.g. the building of cross-border industry networks (3 CIP, EuroTextil, BioEnergyNet, Team22) and cross-industry networks (such as InnoLausitz) will achieve development of the economic region.

Transport Infrastructure

Particular importance is attached to the future development and shaping of the transport infrastructure network in the German border region. If one considers the currently existing transport connections and flows, it becomes clear that different strengths and weaknesses have a lasting effect on the development of the border region. The German border region generally has a better infrastructure provision.

In the transport plan developed for the DPERON region, the following weaknesses for the individual types of transport were discovered:

Weakness of the interregional traffic network

Network	Weakness/Risks
Road network	- incomplete extension of some road sections
	- gaps in the motorway network on the Polish side; important centres (e.g. Gorzów Wlkp.) are not connected
	- absence of efficient North-South-connections on the Polish side; the same today on the German side until the Oder-Lausitz-Route is finished in the region of Brandenburg
	- absence of an efficient East-West-connection, esp. in the Euroregion Pomerania
	- some sections (e.g. A11 to Pomellen) are in a bad state, accumulation of maintenance requirements
	- the wide-meshed network on the Polish side
Railway network	*esp. on the Polish side:*
	- no efficient connection between Berlin and Wrocław
	- important routes are not electrified
	- the safety devices and the controlling systems of the railway routes do not fit to a certain degree
	- when crossing the border, the locomotives have to be changed, except for Diesel engines
	- no common fare union
	- decreasing public transport, esp. in rural regions
Waterways	- access to the Baltic Sea only via the Oder/Odra
	- the infrastructure of the ports and their gates is partly too weak
	- the connections of the important ports to the hinterland (e.g. Szczecin)
	- partly to shallow and narrow water for inland navigation vessels
	- old sluices limit the size of inland navigation vessels
Air traffic	- under certain conditions the extension of the connections to the airport Berlin Brandenburg International (BBI) are not finished in time
	- regional airports will lose their importance when the airport Berlin Brandenburg International (BBI) is extended

Tab. 5: Weaknesses of supra-regional transport networks by transport type. Source: DPERON 2007: 77/78.

Across the border, East-West connections are now reaching the limits of their capacity owing to the greatly increased volume of traffic - particularly on the roads - and rail connections in the peripheral regions are being closed or are increasingly in danger of being closed. A main problem is still the inadequate state of cross-border transport infrastructure: Where there were more than 100 bridges over the Oder and Neisse before the Second World War, there are now less than 20. At some points the border crossings are only now starting to catch up with the greatly increased volume of traffic.

The forecasts assume that the growth in traffic will take place primarily on the roads. In all available development scenarios for large-scale cross-border transport up to 2020 (Federal Environmental Agency) it is clear that the change in road freight will take place in particular in the new member states - increases of 120 to 130% are expected here. In comparison, increases of only 20 to 40% are forecasted in the "old" member states (EU15). Much smaller increases are assumed for rail transport.

In accordance with the regional development trends and the present situation in the border region, the following premises were formulated in the DPERON project for the development of the border region:

- ***Fundamental premises***
  Good quality transport infrastructure of all transport types and good availability of transport are prerequisites for the building of economic co-operation. The existing weaknesses are to be eliminated not only by means of increased investment but also by improved communication, co-ordination and institutional co-operation.
- ***Road***
  Further upgrading of the main road network, especially on the Polish side; removal of the investment backlog for roads, especially in rural areas; further adaptation of the road network and river bridges to the requirements of cross-border traffic
- ***Rail***
  Removal of infrastructural and technical bottlenecks; improvement in accessibility and shortening of travel times, in particular between the German and Polish urban centres; transparent fare structure, optimisation of timetables (including the regional bus network); creation of a cross-border transport association
- ***Domestic shipping***
  Agreement on the upgrading of waterways in the region; investment in port infrastructure (e.g. Szczecin port); improved marketing; connection of ports to road and rail
- ***Air transport***
  Upgrading of Berlin-Brandenburg International airport

In summary it can be said that, in the future, it will be important to use coordinated planning and the joint formulation of priorities to create efficient transport types and systems, and design them in such a manner that the whole border region can benefit from them. For example, frequently used railway lines must be electrified, differing technical requirements in cross-border transport of the two states must be matched to each other, and meshing systems must be developed across borders.

The already existing East-West connections and the partially existing North-South connections are to be extended in a sensible manner. An example of a suitable North-South connection (in the mid- or long term) with linked transport types is the lower course of the river Oder (roughly between Berlin and Szczecin). Here, in the immediate border region, important transfer points for goods (logistical sites) could be created by means of future railports. The Oder or immediately adjacent canals (e.g. the Hohensaaten-Friedrichsthaler waterway) would however need to be upgraded and further infrastructure would be necessary for this.

Essential for the border region is therefore the improvement of **cross-border** transport infrastructure in the short term. The strengthening of economic centres must be the focus of improvement of transport quality. In the field of tourist route connections, the planned "bridge-building over the Oder" and a stronger thematic and spatial networking of tourist routes are primarily to be investigated.

The configuration of the transport infrastructure in the border region must be oriented towards the needs and demands of the employment market, education infrastructure, and also tourism, in accordance with the changed requirements of the centres, peripheral regions, and also cross-border co-operation (Brol 2005, Hamburg Institute of International Economics 2006, Leibniz Institute for Regional Geography et al. 2006). For priority projects, "chronological synchronisation" is vital in order to complement each other, in particular from a cross-border point of view.

The transport infrastructure can however only fulfil the altered requirements if coordination processes already under way in the development region are successfully implemented in a targeted manner in accordance with the DPERON approach, independently of political and regional sensitivities.

Education and Training

Discussion about the education and training of young people has increased in recent years owing to the much noted PISA studies in Germany and Poland. Quality, content, or learning conditions are scrutinised more critically. In the border region, the number of students increased sharply in the course of the transformation process. For example, the proportion of students on the Polish side increased from 8% (1988) to 55% of 19-24-year-olds in 2007 (Wóycicka 2007: 3). The level of education however remained below that of other European countries. The same can be observed in professional training, so that even today a considerable shortage of skilled labour is to be noted in the German-Polish border region. The lack of interest of employees in further training has an exacerbating effect on the Polish side.

The border region, which already suffers from the consequences of demographic change, faces massive problems. Numerous schools already have to be closed or merged due to lack of students. There is also a lack of specialised teachers as a

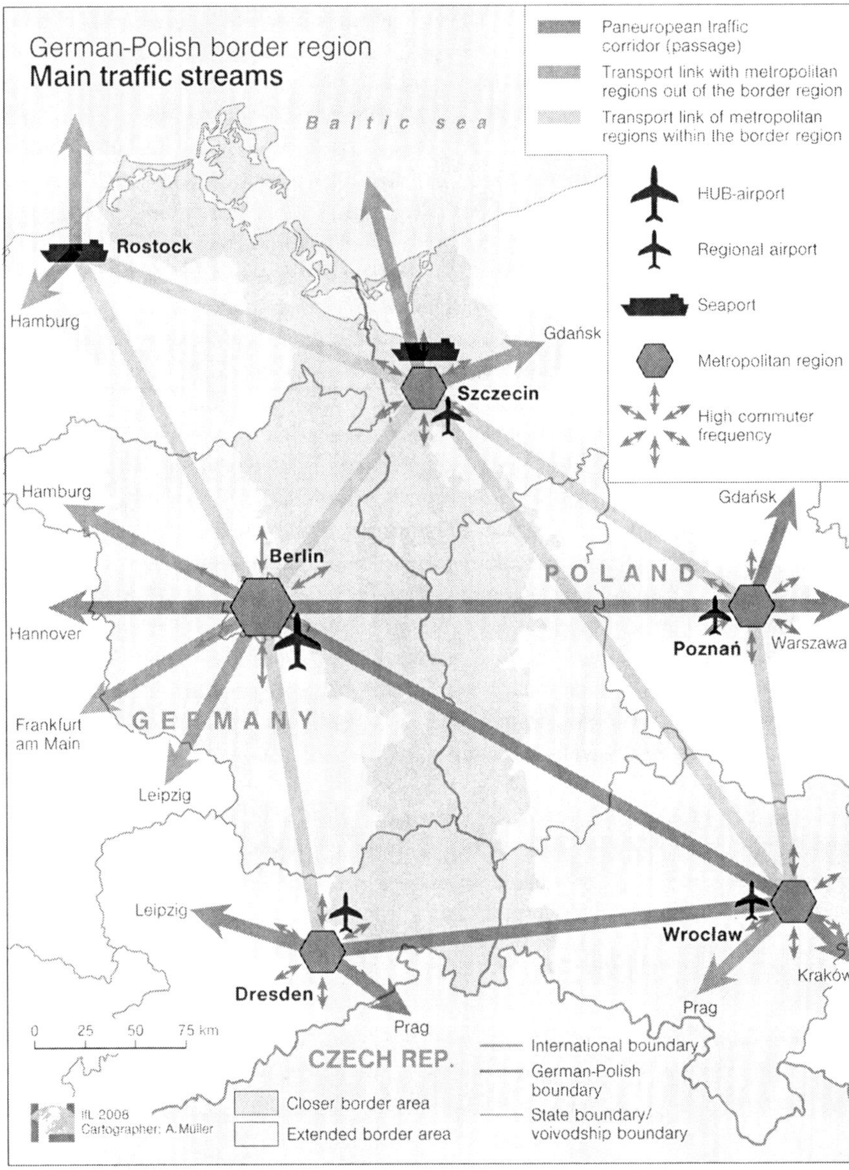

Fig. 7: Main traffic flows in the German-Polish border region in 2020. Source: Hamburg Institute of International Economics 2006.

result of the continuing out-migration from peripheral regions. Polish teaching qualifications are not recognised on the German side and vice versa, owing to differing educational content. Even simply crossing the border in order to visit educational programs in Poland is hampered by increased time, personal and organisational effort, but also by legal regulations.

Fig. 8: Existing education networks in the German-Polish border region . Source: www.lernende-regionen.info.

Starting from these problems and using the traditionally good contacts between schools in the border region, many kinds of networks have been established in the German-Polish border region in recent years (fig. 11, 12), which, each with its particular emphasis, have anchored the subject of "Learning Regions" in German-Polish projects.

In addition to German-Polish language training, the development of common professional training concepts is gaining importance. The following spheres of activity are in particular focus of cross-border co-operations (Learning Regions, 2005):

- Personal development in small and medium-sized companies (SMCs)
- Qualifications for training advisors
- Online training advice
- Euregio Competence Centre - Euroregional exchange for training places
- Multilingual learning materials for daycare centres and guidelines for multilingual daycare centres
- Polish courses for preschool children

Many good ideas of German and Polish schools and children's facilities are however hindered by existing legal framework conditions. In addition, a lower interest of Germans in learning Polish is still to be noted in the border region.

The universities in the border region have, in recent years, reported steadily growing student numbers. This can be traced back to Polish students coming to Germany. Student exchanges are also currently characterised most of all by Polish students studying at German universities in the border region (fig. 13). The

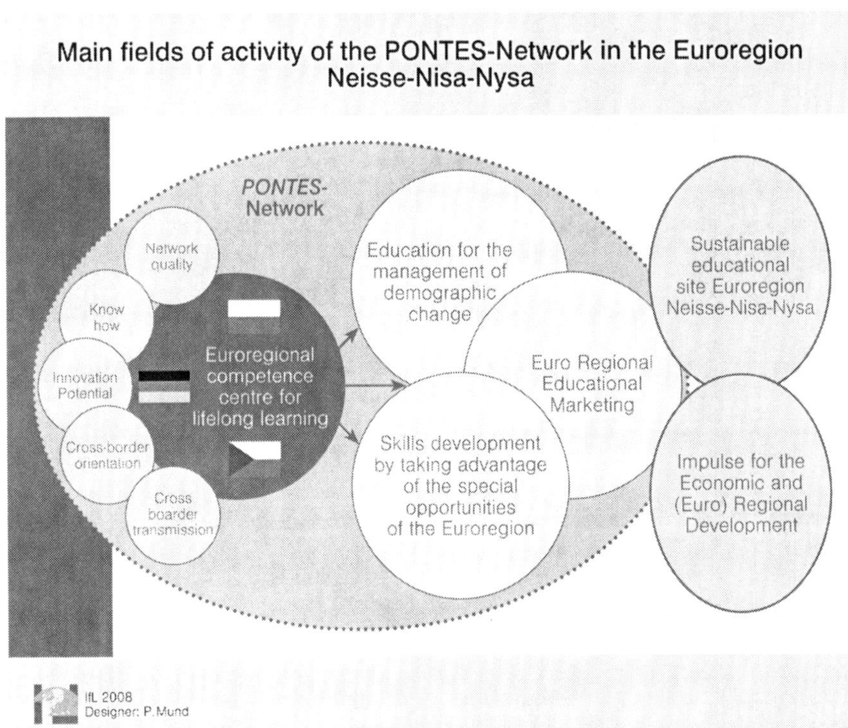

Fig. 9: Priorities of the PONTES network of the Neisse Euroregion. Source: www.pontes-pontes.de/bf/home/netzwerkthemen.html.

universities in the border region are faced with a sharp decline in numbers in the future due to the declining population. In the German border region, the number of students receiving admission to higher education will decline by one-third in 2020, compared to 2003. The same is to be expected for the Polish border region (Hamburg Institute of International Economics Report 262: 61).

Ensuring and developing training and research potential is essential for establishing the border region as a centre of knowledge. This requires a critical mass of scientific personnel and equipment in the individual sectors:

- **Knowledge clusters** (universities; research institutions; technology transfer points; technology and new business centres)
- **Innovative economic sectors** (indicators of the innovative potential of the economic area)
- **Knowledge brokers** (special/further schools; libraries; archives; museums with research tasks; in addition theatres, adult education centres)

In this case, the German-Polish border region has, in certain locations, good starting conditions, in particular due to already existing networks. An important aspect of German-Polish educational and scientific relationships are the university partnerships. Various German-language educational programs in Poland, the European University "Viadrina" in Frankfurt (Oder) refounded in 1991, and the program of the "Neisse University[2]" or the "International Graduate School of Zittau" intensify mutual scientific and cultural exchange. Language barriers, a lack of recognition of educational achievements and qualifications in certain places, but also financial questions prevent an expansion of cross-border study programs, however.

On the basis of the agreement on scientific and technological co-operation of 1989, co-operation has also taken place, promoted by the Federal Ministry of Education and Research and the German Research Foundation (including two German-Polish graduate schools).

Existing good approaches to co-operation, e.g. in the "City Networks of Science" initiative in the Neisse Euroregion (Bergfeld and Groß 2005), have not grown into new, innovative education plans for specialisation and cross-border co-operation.

The existing, also cross-border, active knowledge bases (universities, technology centres and transfer points) are opposed by a series of development obstacles (Bergfeld and Groß 2005: 25ff.), which make development in the border region more difficult, such as

- a lower proportion of R & D employees compared to employees paying social insurance contributions (< 7 %);
- a low number of patent applications;
- R & D financing of universities by regional business of only about one tenth of budget;

2 Joint courses of study of Zittau/Görlitz University, Liberec TU and Wroclaw/Jelenia Gora University in English.

- a lower proportion of companies with their own R & D capacity (German border region: approximately one third of companies);
- a broad range of education options but also of technological fields, only rudimentary concentration on core competencies;
- a low level of research co-operation between universities in the border region and with non-university research institutes.

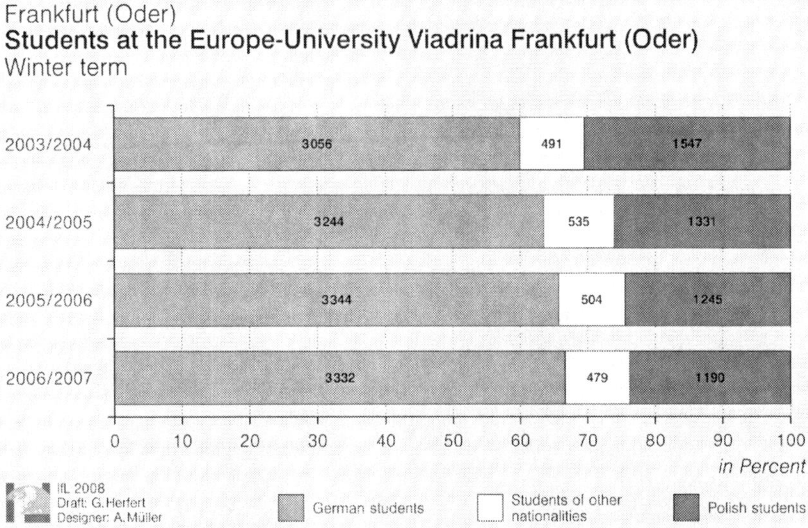

Figure 10: Background of students at the European University Viadrina Frankfurt/Oder. Source: http://www.euv-frankfurto.de.

In the coming years, prerequisites for establishing the border region as a region of learning and knowledge are therefore:

- cross-border information transfer of R & D achievements at a higher level;
- development of cross-border co-operation of universities - stronger specialisation and co-operation and expansion of joint courses of education with unique characteristics;
- co-operation of innovative companies and an increase in the efficiency of business networks;
- national and international involvement of the regional economy, taking into account transregional and cross-border co-operation;
- cross-border trainee contacts and training;
- improvement of foreign language skills.

9.3 Quo vadis, German-Polish Border Region?

With the expansion of the EU to the East in 2004 and Poland's entry to the Schengen Agreement in 2007, new development opportunities arose for the German-Polish border region, even if a significant barrier in the form of continued access restrictions to the German employment market still exists.

How can we exploit these development opportunities? Which developments should be expected or aimed at on the basis of the existing framework conditions? As one cannot assume that the border region as a cohesive economic area will develop independently, a strategy for integrating the border region into larger national or international development regions is necessary. To date, this process has been restricted to linking individual nodes of growth in the border region to other European centres. With the initiative of the eastern German states for spatial development in the Baltic-Adriatic Corridor (Berlin Declaration 2007), the intention is that the particular development potential along the new internal borders of the EU, addressed in the EU's cohesion report, will be utilised more intensively (fig. 14). The objective of the initiative is the creation of an attractive North-South transport infrastructure, in particular the upgrading of the trans-European transport networks, in order to noticeably reduce rail travel times between the metropolitan regions, build multi-modal logistical chains, and improve the quality of sites for the location of transport-related and other innovative industry sectors, among other goals (Berlin Declaration 2007). Even if we are still only at the start of potential development with this initiative for the Baltic-Adriatic Corridor, it still forms an opportunity within the framework of the transition to a knowledge-based society to develop new spatial structures as alternatives to the established main transport axes and the European core region between London and Milan (Kujath 2007).

The linking to form an international development corridor presupposes however that the growth poles of the border region position themselves as attractive participants in international competition. With regard to competition and the foreseeable demographic conditions, strategies for a process of spatial concentration in the border region have repeatedly been produced in recent years (Hamburg Institute of International Economics 2006; Leibniz Institute for Regional Geography et al. 2006), which indicate a stronger reference to the German growth poles of Dresden, Berlin, and Rostock and the Polish growth poles of Posznan, Wroclaw, and Szczecin. These centres already form the economic core areas of the border region, without the positive development of which the border region cannot exploit its opportunities in the future. A policy of promoting such spatial concentration processes logically involves an intensification of the depletion processes in other parts of the border region. At present, there are already significant migration movements out of the peripheral areas. Orienting these more strongly towards the regional centres of the border region would be a positive objective of the promotion policy. This might also counteract the further weakening of small and medium-sized towns.

References

BAMF (2005): Migrationsbericht 2005 des Bundesamtes für Migration und Flüchtlinge. Nürnberg.

BBR – Bundesamt für Bauwesen und Raumordnung (ed.) (2006): Raumordnungsprognose 2020/2050 (= Report, No. 21). Bonn.

Berlin Declaration (2007): Berliner Erklärung zur Raumentwicklung im Ostsee - Adria - Entwicklungskorridor vom 30. November 2007 http://stadtentwicklung.berlin.de/aktuell/pressebox/includes/docs/doc156.pdf

BMVBS – Bundesministerium für Verkehr, Bau und Stadtentwicklung (2006): Leitbilder und Handlungsstrategien für die Raumordnung in Deutschland. Verabschiedet von der MKRO (Ministerkonferenz für Raumordnung). Informationen zur Raumentwicklung, No. 11/12, 701-723.

Brol R. (2005): Raumentwicklung der deutschpolnischen Grenzregion. In: Neumann, I. (ed.), Die Zukunft des deutschpolnischen Grenzraums gestalten. Szenarien für die Europastadt Görlitz/Zgorzelec im Zuge der EU-Erweiterung. Dresden. 115-124.

Glorius B. (2007): Transnationale Perspektiven. Eine Studie zur Migration zwischen Polen und Deutschland. Bielefeld.

Hamburgisches Welt-Wirtschafts-Archiv (HWWA) & Hamburg Institute of International Economics (2006): Der deutschpolnische Grenzraum im Jahre 2020 – Entwicklungsszenario und Handlungsempfehlungen. Hamburg.

Herfert G. (2007): Regionale Polarisierung in der demographischen Entwicklung in Ostdeutschland. In: Raumforschung und Raumordnung 65, No. 5/6, 435-455.

Herfert G., Lentz S. (2006): New Spatial Patterns of Population Development as a Factor in Restructuring Eastern Germany. In: Lentz, S. (ed.): Restructuring Eastern Germany. Berlin, Heidelberg, New York: Springer, 91-106 (German Annual of Spatial Research and Policy).

Hoorn A. (2006): Wahrnehmungs- und Handlungsmuster an der deutsch-polnischer Grenze: Ergebnisse einer Untersuchung unter Jugendlichen in Guben-Gubin. In: Europa regional, No. 1, 22-32.

IfL (Leibniz-Institut für Länderkunde) et al. (2006): Konkretisierung der Leitbilder für den erweiterten deutsch-polnischen Grenzraum. Vorstudie zum deutschen Teilraum. Final report, May 2006 (unpublished).

Jurczek P. (2002): Cross-Border Cooperation in the German-Czech-Polish Border Region at the Turn of the Century. In: Journal of Borderland Studies 17, No. 2, 97-104.

Kaczmarek T., Stryjakiewicz T. (2006): Grenzüberschreitende Entwicklung und Kooperation im deutschpolnischen Grenzraum aus polnischer Sicht. In: Europa Regional 14, No. 2, 61-70.

Konicz T. (2007): Der große Aderlass. http://www.heise.de/tp/r4/artikel/25/25241/1.html

Krätke S., Borst R. (2004): Chancen der EU-Erweiterung: Perspektiven der deutsch-polnischen Unternehmensbeziehungen am Beispiel der Region Berlin-Brandenburg. In: Geographische Zeitschrift 92, No. 1/2, 93-116.

Kujath H. J. (2007): Raumwirtschaftliche Voraussetzungen für die Entwicklung des Ostsee-Adria-Entwicklungskorridors. Vortrag auf der Auftaktkonferenz zum Ostsse-Adria-Entwicklungskorridor http://gl.berlin-brandenburg.de/imperia/md/content/bb-gl/europ_raumentwicklung/1_kujath.pdf

Maack K. et al. (2005): Wachstumspol Stettin und Auswirkungen auf die Entwicklung der deutsch-polnischen Grenzregion. Edition der Hans-Böckler-Stiftung, No. 132. Düsseldorf. http://www.wilke-maack.de/service/stettin_studie_kurz.pdf

Oschlies W. (1983): Polnische „gastarbeiterzy" in der DDR. Rechtsgrundlagen und Alltag. In: Deutschland-Archiv, Zeitschrift für Fragen der DDR- und Deutschlandpolitik 16, No. 10, 1084-1091.

Rutowska M. (1996): Die regionale Zusammenarbeit in der deutschpolnischen Grenzregion in den Jahren 1945 -1989. In: Schultz H., Nothnagel A.: Grenze der Hoffnung. Geschichte und Perspektiven der Grenzregion an der Oder. Frankfurter Studien zur Grenzregion. Potsdam. 42-48.

Sinz M. (2006): Die neuen Leitbilder der Raumentwicklung – Anmerkungen zu einem politischen Diskurs. Informationen zur Raumentwicklung, No. 11/12, 605-612.

10 Study of Spatial Developments in the Polish-German Border Region

Grzegorz Węcławowicz, Marek Degórski, Tomasz Komornicki, Jerzy Bański, Przemysław Śleszyński, Marek Więckowski

Introduction

This contribution is the authors' version of a document commissioned by the Polish Ministry of Infrastructure in 2005 titled "The Guideline Study of Spatial Development alongside the Polish-German Border". This work has been elaborated by a team from the Institute of Geography and Spatial Organization of the Polish Academy of Sciences in Warsaw as well as by invited experts from outside[1]. The study carried out by this team reflects the new socio-political and economic situation following Poland's EU accession, as well as the highly dynamic nature of the social and economic processes ongoing in the country's western borderland. The situation generates new opportunities and challenges for regional development policy. In particular, this concerns newly emerged prospects for establishing favourable conditions supporting cohesion and economic activation of the western part of the Polish borderland. This may be accomplished through such measures as, for example, launching investment projects within the framework of structural funds, decreasing barriers in cross-border traffic, as well as facilitating the free flow of people, goods, and services.

The study is an attempt to summarize and synthesize the relevant research, as well as the planning level and strategic studies compiled to date. The paper includes the results of the spatial analyses that describe and explain the current situation and the structure of spatial organization of the Polish part of the western borderland, as well as present the general directions of development in the long-term perspective. These analyses may become a basis for determining the internal and external opportunities and threats.

The problems addressed in the studies covered reflect the requirements of the Ministry to a large degree, extended by the author team's concerns on issues such as: the natural environment and protected areas, flood protection, the settlement system, demographic processes, the selected problems of economic and social

1. "The Guideline study of Spatial Development alongside the Polish-German Border"
Key experts: Grzegorz Węcławowicz (Team Leader),
Marek Degórski, Janusz Korzeń, Tomasz Komornicki,
Other experts: Jerzy Bański, Juliusz Korzeń, Roman Soja, Przemysław Śleszyński, Marek Więckowski.

development, rural space and agriculture, trans-boundary linkages, transport and technical infrastructure, tourist infrastructure, and regional development.

The whole document (Guideline ...) constituted the selection of 300 detailed analytical maps together with synthetic maps covering the study area at 1:200,000 scale, a statistical appendix, and the list of developmental projects being in the process of realization, in the preparation phase, or planned in regard to the *gminas* (basic local self-government level) situated within the borderland. This chapter contains the most important ideas, summary and conclusions.

The research studies, gathering the materials and opinions, as well as field consultations, unexpectedly for our team, attracted much attention of the public opinion and local authorities on all levels within the borderland *voivodships* (regional level). An active part in the devising of the analysis was also played by the numerous participants of regional - and local - level consultation sessions

Fig. 1: Study area.

held in Wrocław, Zgorzelec, Słubice, Zielona Góra, Szczecin and Poznań. Also of major cognitive importance was questionnaire-based research carried out regarding matters of planning and physical development in the gminas located adjacent to the border. This was backed up by field research carried out by the author team.

Public Reactions

It seems peculiar that during the time the project was carried out, it actually failed to generate any considerable public interest. Only after finishing the research studies and publishing their results, some local authorities, especially on the part of Lubuskie Voivodship, began to impose their particularistic interests, quite often contrary to nation-wide or to other voivodships' visions (e.g. S-3 expressway). Particularly, the confusion caused by the local press was of considerable significance. The Gorzów edition of the Gazeta Wyborcza of 29 August 2005 reported that the study is, in many places, contradictory to regional strategies of the Lubuskie, Dolnośląskie and Zachodniopomorskie voivodships. The press included a scheme of road linkages different to the original scheme proposed by the experts working on the study.

The basic problem was a misunderstanding of the aim of the paper. Since the results of our research study showed the current state of affairs and offered our opinions, which originally were not supposed to be the passive summary of the concepts contained in the plans and regional strategies developed so far. The comments concerned many issues which, according to our team, were devoid of creativity or were principally mistaken. These contents were left unnoticed or openly criticized. The best example was the issue found in the planning documents of the Lubuskie Voivodship concerning the hardly realizable concept of the duopoly Zielona Góra-Gorzów Wielkopolski. Criticizing this concept and undermining its validity led to attempts to negate the value of the whole study.

Regional authorities of the Lubuskie Voivodship claimed unjustly that the results of our research study posed a threat to the plans to construct expressways, especially the S-3. At the same time, the fact of preparing the documents led Bogusław Liberadzki, a member of the European Parliament, to raise alarmist concerns over the supposed delay of the Ministry of Infrastructure sending documents in regard to financing the road investments to the European Commission. It is worth noticing that this artificially created problem paradoxically had a positive effect, as it led to some integration of the voivodship – regional community in defence of their own interests.

As a result, the team's effort put into the research studies was undermined – in spite of the fact that the study was carried out according to the ministry's guidelines. In particular, the team objected to altering the results – during consultations, and especially in the process of accepting the final draft of the paper.

After conducting analyses and, in particular, after arranging extensive field consultations and discussions with local authorities at the different levels as well as with regional planners, the authors of the study came to the conclusion that there is more than one interpretation of the obtained research results. Unquestionably, it was a mistake not to take into consideration political background, local and regional groups of interests, insufficient access to information on the part of the main recipients of the study, and information chaos created by the local media.

The authors assumed that the best means of verification of the results (not only from the scientific point of view) would be to publish them. Since the full accessibility of the study through revealing all the analytical materials and research results may allow a broader range of people (not only planners and scholars) their own analysis and interpretation. Due to the breadth of the analyses, part of the materials (including the aforementioned cartographic appendix) was published in electronic form on the included CD ROM[2].

10.1 Assumptions and Objectives of the Study

The primary objectives of the analysis have been to ensure spatial, economic, and social cohesion of the Polish-German borderland and the rest of Poland, plus the eastern parts of Germany.

The second objective was an attempt to point out means in which, and directions through which, an area regarded as a peripheral, unfavourable location might domestically (in Polish terms) take full advantage of what is seen as a favourable location in the wider European context.

If the aforementioned, negative characteristics partial to a peripheral location are to be dispensed with, it will be necessary to ensure the integration of economic and social activity, as well as diffusion of innovation across the border, and a drawing of full economic and social benefits from the fact that these border areas are close to the rapidly-integrating metropolitan areas of Szczecin, Poznań and Wrocław (on the Polish side) and Berlin (on the German side). It seems to be obvious that the aforementioned objectives can be accomplished only when improving the level and quality of life, as well as stimulating the economic growth and competitiveness within the whole area of the Polish-German borderland.

The high level of detail of the spatial analyses involved is intended to provide for the study's use in the national, regional, and local systems of spatial and strategic planning in Poland, as well as the regional policy of the EU as a whole. It is assumed that the document will provide a basis for the identification of key directions regarding development and spatial organization in the western borderland, these being agreed upon at the interregional (voivodship) level and internationally (between Poland and Germany). This will then be of value when it comes to the compilation of reports on the situation regarding spatial organization and physical

2. Węcławowicz G., Degórski M., Komornicki T., Korzeń Ja., Bański J., Korzeń Ju., Soja R., Śleszyński P., Więckowski M., (2006), Studia nad przestrzennym zagospodarowaniem obszaru wzdłuż granicy polsko-niemieckiej. Prace Geograficzne nr 207, IG i PZ PAN, Warszawa. (Studies on spatial development of the polish-german border region. Geographical Studies No. 207).

development, work to update the strategic studies and plans at voivodship level, and necessary updating of spatial organization concepts and the National Development Plan for the years 2007-2013 (the study was prepared in 2005 – the work on that part of the above-mentioned documents was still ongoing). The study may also prove suitable as the National Development Plan for the years 2007-2013 is being developed, as well as representing one element in the preparation of the Polish voivodships for their future negotiations with the European Commission over the Regional Operational Programmes for the years 2007-2013.

The content of the study reflects the analyses and studies carried out on the physical development plans and development strategies (or update thereof) in force at voivodship level, Poland's draft National Strategy for Regional Development 2007-2013, and the country's National Development Plan for the years 2007-2013, as well as the Lisbon Strategy at regional level. Efforts have been made to ensure that provisions are in line with what has been adopted in the aforementioned documents, and that the programming of development is adjusted to the EU model.

The materials are an element in discussion regarding the shaping of a future vision as far as the functioning of Poland's western borderland is concerned – a vision that should clearly define the opportunities for, and interests of, both the local and regional authorities and the business entities active in Poland, as well as offering the articulation of these interests within the framework of the supraregional (supra-voivodship level) and cross-border structures of the integrating European Union.

As far as spatial organization of the Polish-German borderland is concerned, the various studies and expert opinions generated hitherto (mainly by researchers and planners on the German side in the 1990s) brought into existence a number of concepts, the implementation of which could have led to intensification of the (economically and socially) unbalanced development of areas lying either side of the border, and hence to a marginalization of the Polish part of the borderland. In a nutshell, these spatial concepts were markedly unfavourable for Poland. In the light of this, the present study aimed, as one of its tasks, at some restoration of "balance" as far as the Polish-German borderland is concerned, by way of an indication of alternative directions that its development and functioning may take, in an understanding that the primary goal is to limit – and, in longer term, to eradicate – the disparities existing at present. Such a reduction of disparities is also in the interests of the business entities operating on the German side, as it would lead to a more prominent position for the Polish-German border macro region as a whole (which has a chance to become integrated spatially and functionally in the future) within the ranking of regions of integrating Europe.

The study has been devised with the determining of the situation and trends regarding the management of the Polish part of the Polish-German borderland in mind. A supplementary study of a similar nature should be provided by separate work concerning the German borderland area. The two studies taken together will then serve as a basis for a synthetic study regarding the management of the region.

10.2 Spatial Delimitation of the Study

Methodology adopted for the study results from the range accepted within the framework of the commissioned task, as well as from arrangements worked out by the Polish-German Committee concerning Spatial Management. The subject scope was extended by the author's team with issues such as, for example, demographic issues, problems of rural areas, as well as trade and services development.

Determining the research field and the range of study works on the Polish-German borderland area brought various solutions so far; unsatisfactory, however, for all parties interested. This area previously covered the four former voivodships (Szczecińskie, Gorzowskie, Zielonogórskie and Jeleniogórskie). In 1999, after the administrative reform of Poland and changes in geopolitical conditions, this area was extended to include the Poznań and Wrocław agglomerations.

The area accepted for research and analyses is related in spatial range, as it was mentioned in the Updating of the Guideline Study of the Spatial Organization of the Area along the Polish-German Border of 2002. This area covers the entire Dolnośląskie and Lubuskie voivodships, the western part of Zachodniopomorskie voivodship, as well as the four *powiats* (including Poznań) belonging to Wielkopolskie voivodship (fig. 1). This delimitation of the area, as indicated by the Ministry of Infrastructure as client, was influenced by such factors as, for example, the necessity to respect the previously accepted mutual agreements, and also the need for concentrating on a few geographically defined axes and developmental areas involving the space of the whole macro region of the western borderland.

Attention should be paid to spatial diversification of the borderland area as such. Thus, the delimited area of research consists of a few historically shaped regions. An attempt to create a new borderland region is an idea contradictory to consolidation trends in the regional perspective, because it artificially delimits the boundary of the region, the only unifying force of which is a group of phenomena related to geographical location. The aim of regional policy rather ought to be shaping the regions characterized by clear economic relations, especially by economic and social cohesion. The historical region of Silesia may serve as an example on the Polish side, the boundaries of which have been largely unchanged for 900 years. The historical cohesion sources can be strengthened by regional policy. The development of regional cohesion of Pomorze Zachodnie requires much greater resources, and in the case of Lubuskie Region, the regional cohesion actually needs to be created practically from the very beginning.

Delimitation of the borderland area on the basis of the range of influence and economic relations distinguished by specific characteristics could be of basic significance, however this always remains controversial. The research studies of that type should answer, above all, the question whether an actual borderland or transboundary region exists as a real entity, or whether it exists as a political entity, i.e. as a delimitated territory, where regional policy is conducted by the state.

Conducting detailed and specific research studies is a precondition for solid and fair delimitation of the borderland region. Numerous methods of spatial delimitation applied in geography and regional research exist which compel to take basic decisions with regard to delimitation objectives and criteria applied.

In case of the state's regional policy, the basic objective of borderland regions is to determine how far inside of the Polish territory socio-economic influences generated from German territory and the rest of the EU can be observed. Within this context, it is important to answer the question whether these influences are positive or negative. As a rule, a considerably larger research area tends to be covered in order to capture the spatial structure inside of the borderland region. It appears that this course of action was adopted by the client in regard to The Guideline Study of Spatial Development Alongside the Polish-German Border. It was assumed that the most important phenomena typical of the borderland region would be captured to a much larger extent. So, by making such assumptions, it would be necessary to include the remaining part of the Zachodniopomorskie voivodship, Opolskie voivodship (having particularly close relations with Germany), as well as sub-regions (NUTS-3) situated in the western part of the Wielkopolskie voivodship into the future research study. Inclusion of these latter mentioned units was insisted by, for example, authorities of Wolsztyn Gmina; despite the fact that this gmina is located within boundaries of Szprewa-Nysa-Bóbr Euroregion, it was excluded from the research borderland area due to a top-down decision. The range of direct influence of the border may be characterized by a distance function; however, its measurement must be carried out in accordance with criteria based on a time accessibility index. In this case, areas immediately adjacent to the border and alongside the transport corridors running towards the border are of particular importance.

It may be assumed that the power of influence and economic plus social links is decreasing along with increasing physical distance and time accessibility. The minimum area of detailed research concerning the direct influence on the borderland must at least take the belt of borderland sub-regions into account.

Acknowledging of two delimitation scales and analyses of the borderland region, namely local scale (gminas immediately adjacent to the border) and supra-regional scale (neighbouring voivodships or parts of the voivodships), reflects two types of socio-economic phenomena taking place in the western part of Poland. For local and neighbourly links conditioning common day-to-day existence within the borderland zone are of different nature, as well as another way economic and social processes are being developed deep inside the neighbouring countries and affecting the country as a whole or even all of Central Europe.

10.3 Diagnosis

Analysis of the planning- and strategic-level documentation for the Polish-German borderland that is available at local, regional, and national levels has been combined with the results of diagnostic analyses and the knowledge available to the author team from their study of the literature and from numerous consultations in the field, to allow for a presentation in the following tables listing the most important opportunities, barriers, and threats. In line with the objectives of the Guideline Study of Spatial Development Alongside the Polish-German Border, the list has taken on a hierarchical structure. It may be augmented by new elements in the future.

Subject	Opportunities	Barriers and threats
1. The natural environment and protected areas	• High level of forestation and abundance of surface waters (in Lubuskie and Zachodniopomorskie voivodships) • Very diverse landscape • Clean state of most environmental components (other than waters) • Valuable natural features and significant geo- and biodiversity • Cohesive system of protected areas in the Oder, Warta and Noteć Valleys	• Limited resources of surface and subsurface water (Dolnośląskie voivodship) • Fragmentary system of protected areas over most of the borderland area • Disorganised systems for the management of water and wastewater, as well as municipal and industrial waste • Proposals to create new protected areas not widely accepted among local communities or businesses • Absence to date of comprehensive environment-friendly ways of linking up environmental protection and economic development
2. Flood protection	• Multifunctional retention reservoirs (serving power generation, navigation, water supply and other recreation) • Flood warning system in place to prevent catastrophic events • Facilitated use of water, especially regarding navigation and wastewater discharges	• Pressure regarding the economic use of valley areas within flood banks • Flood-induced losses, especially in cities and regarding hydrotechnical infrastructure • Inadequate funding, high costs, and lengthy delays regarding investment in hydrotechnical infrastructure • High level of uncertainty regarding the ultimate effectiveness of pro-environmental investments • No agreed Polish-German action programme for flood protection
3. Problems with the functioning of the settlement system in the Polish-German borderland	• Dense, well-developed settlement network (in Dolnośląskie and Zachodniopomorskie voivodships) • Integrated central areas of borderland cities and revitalisation of other town centres • Development of service, residential and commercial functions • Favourable geographical location	• Good access to metropolitan areas lacking • Hierarchy of settlement system unclear • Single-function economies in some urban areas (Dolnośląskie voivodship) • Dispersed built-up areas and increased operational costs regarding suburban infrastructure • Lack of socioeconomic and spatial cohesion of settlement system
4. Demographic processes	• Higher demographic dynamic on Polish side • Greater degree of population in pre-productive age • High spatial mobility of population • Development of higher education in regional centres	• Outflow of population from periphery • Intensive and uncontrolled influx of people into suburban areas • Ageing of population in largest cities and depopulation areas • Outflows of the best-educated and most entrepreneurial from rural areas and small towns • Increasing disparities in population structure by age and sex

5. Issues of economic and social development	• Large workforce in urban areas • Urban areas looking increasingly attractive as places to locate capital • Metropolitan areas have potential to activate areas adjacent to settlement • Mineral raw materials of prime importance to economy (brown coal, copper) • Foreign capital being utilised in economic activation	• Severe unemployment and social exclusion problems; • Low level of economic and social activity in some areas • High level of burdening of the economically active and low level of professional activity • Limited investment expenditures in local authority budgets • Limited educational attainments and poor school results except for agglomerations • Local authorities show lack of continuity and failure to focus on selected pro-development activity • Closures and staff reallocations associated with transfer of functions from western to eastern borders • Trend towards equalisation of prices for products and services
6. Issues relating to rural space and agriculture	• Multifunctionality of most rural areas • Potential of productive agricultural space to generate organic food • Agrotourism in Lakeland and upland/mountain areas • Agrarian structure favouring enlargement of farms • Metropolitan areas as markets for food products	• Rural population poorly qualified • Agroecological conditions poor in some areas • Borderland farms underinvested in, while infrastructure of former State Farms de-capitalises • Closure of State Farms resulted in poverty, social exclusion and extensification of farm output • Hindered transport access of country-dwellers to local service centres
7. Functional linkage on the national and transboundary scales	• International linkage of a social and economic nature growing in strength • Local economy geared towards export • Heavy border traffic and growing German demand for Polish services • Concentration of commerce and services around certain border crossings • Concentration of transit traffic along main arteries • Decentralization of passenger flights • Proper acknowledgement of consequences of entry into Schengen Agreement	• Limited diversification of foreign economic partners (Germany prevails) • Limited level of processing of exported goods (in northern part of study area) • Delays with transport developments (motorways and main roads)
8. Transport and technical infrastructure	• Transport accessibility of EU's main economic centres • Good level of equipment of border in border crossings, *i.a.* those for tourist traffic of local significance • Relatively large share of higher-order investments in communications being made in Poland • Dense road system	• Weak transport linkages between northern and southern parts and central Poland • Inadequate transport linkages between borderland and Western Polish metropolises of Szczecin, Poznań and Wrocław • Limited no. of exits (junctions) on built motorways generating "tunnel effect" (especially on A2) • Shortfalls in roads leading to border crossings (gaps in network) • Competition Szczecin/Świnoujście port complex faces from Hamburg and Rotterdam and growing importance of road freight • Declining importance and suitability of Oder as waterway, dangers of navigating into Germany via the Stara Odra (Old Oder) channel • Separation of Świnoujście's communications system from its surroundings • Lack of institutional options in support of intermodal freight

9. Tourist infrastructure – centres and development	• Developing supply and demand regarding domestic and foreign tourist services • Outstanding natural and cultural features plus attractive landscape as foundation for further development of tourism • Particularly valuable features for waterborne tourism (inland and by sea) • Special conditions for health and spa-based tourism • Increased interest in buying land and second homes	• Low-standard accommodation base and supporting infrastructure • Difficulties with returning to a tradition of tourism and region's lack of "identity" among potential clients • Possible downgrading of some border areas to servicing of transit traffic only • Short season and overburdening of some centres • "Takeover" of potential clients for borderland tourist services by more efficient and better-organised agencies on German side • Competition from alternative tourist areas close to borderland area • Inadequacy of pricing in relation to quality of services offered • Inadequacy of marketing of tourist centres and regions • Poor accessibility
10. Regional development	• Borderland developing logistical and communications functions that meet needs of European economy • Influx of foreign capital and assistance funding • Development of educational and research centres (Poznań, Wrocław, Szczecin) • Functional diversification of borderland area • Taking on of joint actions in the name of development by the German side	• Decapitalising infrastructure and continued underinvestment in border areas • Economic depression in some towns and areas formerly possessing State Farms • Marginalisation of areas of more limited accessibility • Limitations on development imposed by *Natura 2000* • Deteriorating commercial conditions for some of region's products (mining and agricultural) • Spatial separation of competences as regards administration and management (especially in the cases of Gorzów Wielkopolski and Zielona Góra) • Impractically theoretical nature of strategic planning documents for practitioners (a surfeit of "axes of development", "corridors of international significance", "gateway cities", etc.) • Passivity of regional-scale spatial planning with shortage of new ideas and overestimation of transit and tourist functions that merely repeat initiatives first put forward in era of central planning • A lack of cohesiveness with necessarily-updated studies at local authority level and local plans

Tab. 1: Opportunities, barriers, and threats in the regional development of the Polish-German borderland. Source: Węcławowicz G., Degórski M., Komornicki T., Korzeń Ja., Bański J., Korzeń Ju., Soja R., Śleszyński P., Więckowski M. (2006) Studia nad przestrzennym zagospodarowaniem obszaru wzdłuż granicy polsko-niemieckiej. Prace Geograficzne nr. 207, IG i PZ PAN, Warszawa. (Studies on spatial development of the polish-german border region. Geographical Studies No. 207, Warsaw).

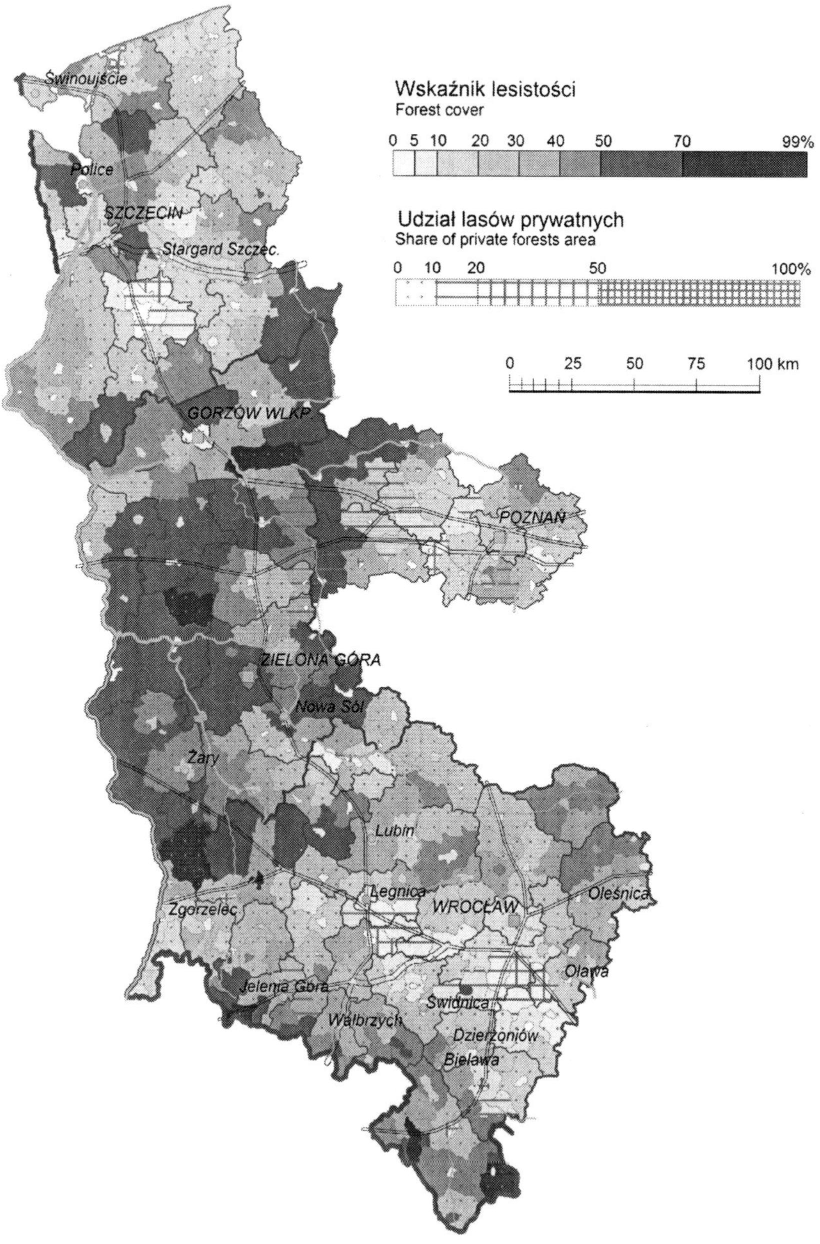

Fig. 2: The natural environment and protected areas.

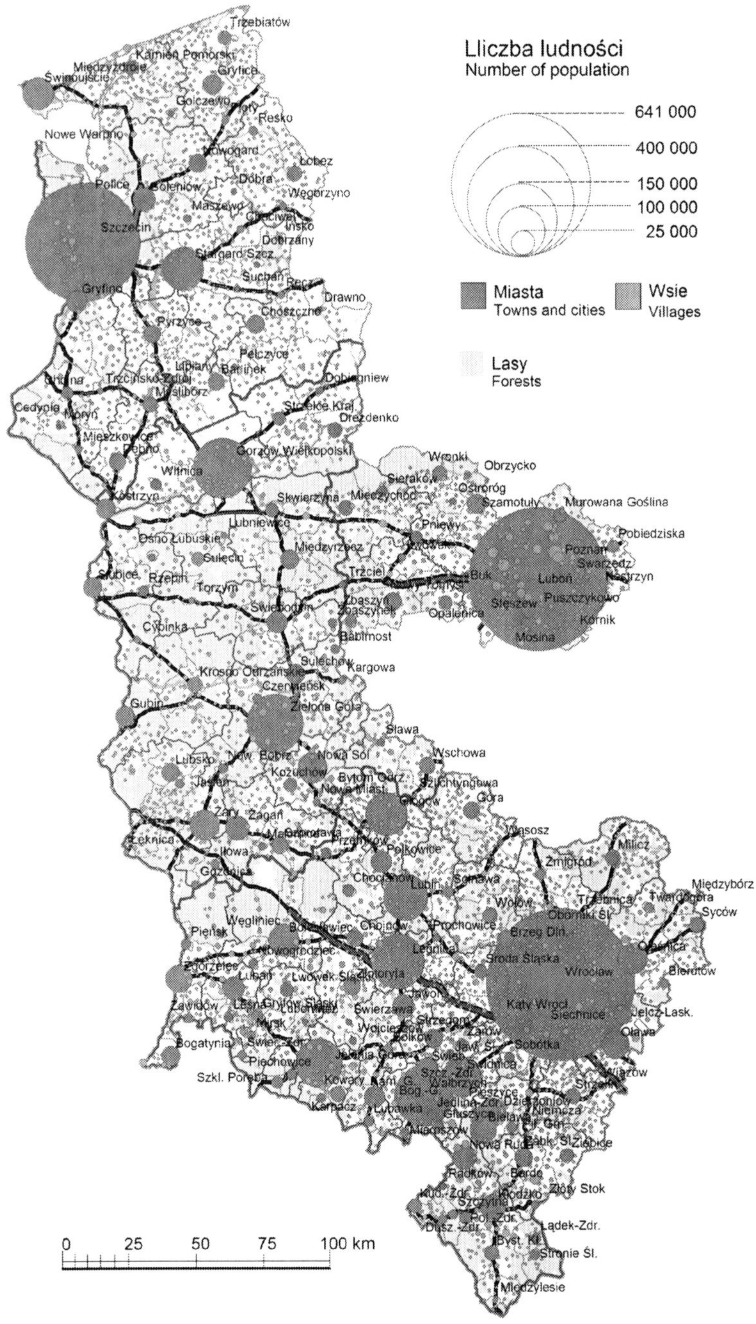

Fig. 3: Number of population.

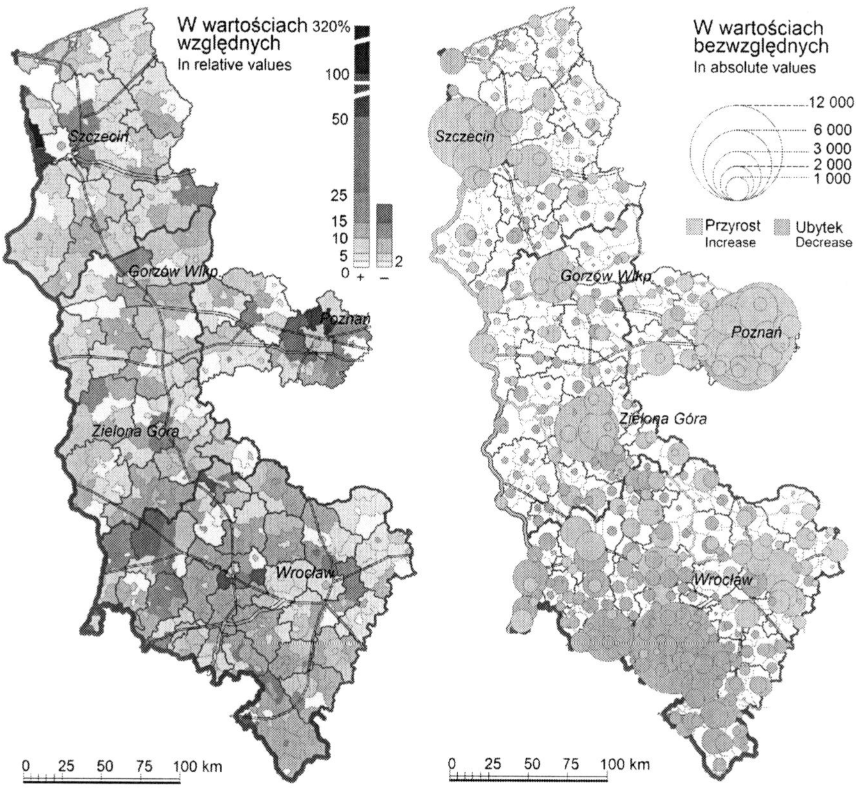

Fig. 4: Demographic processes.

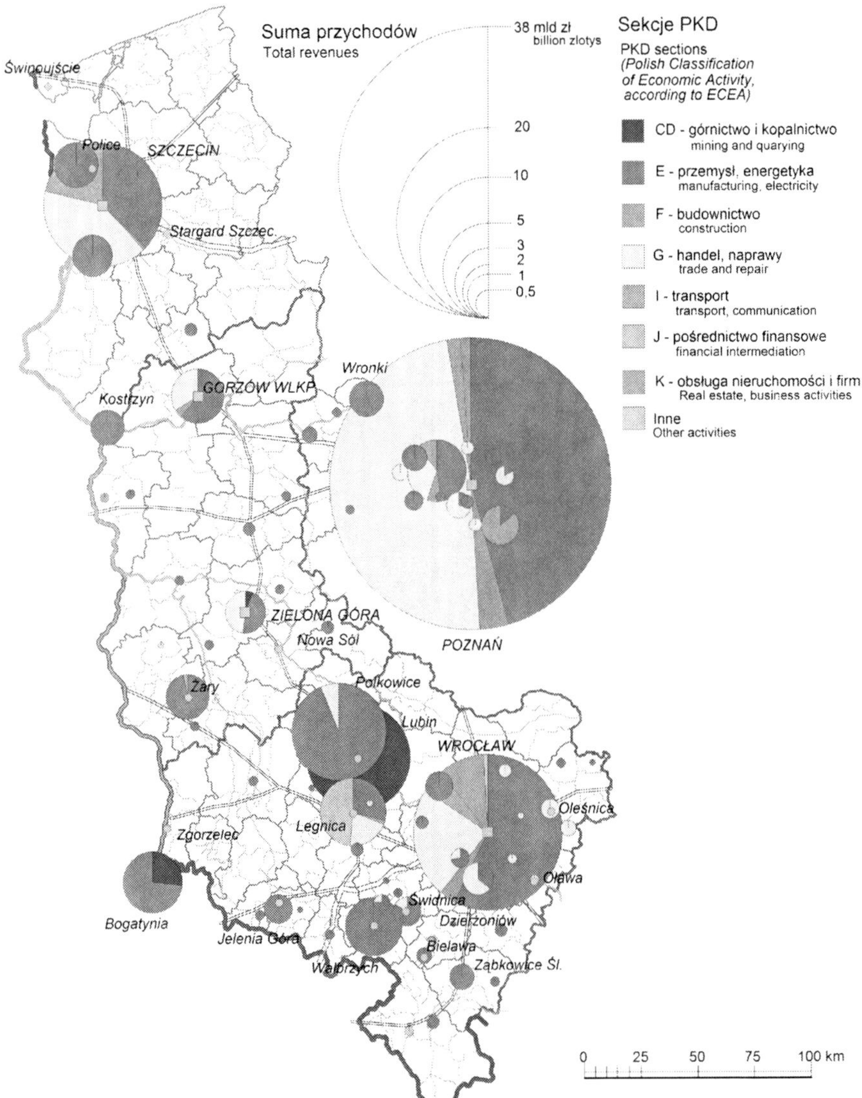

Fig. 5: Economic development.

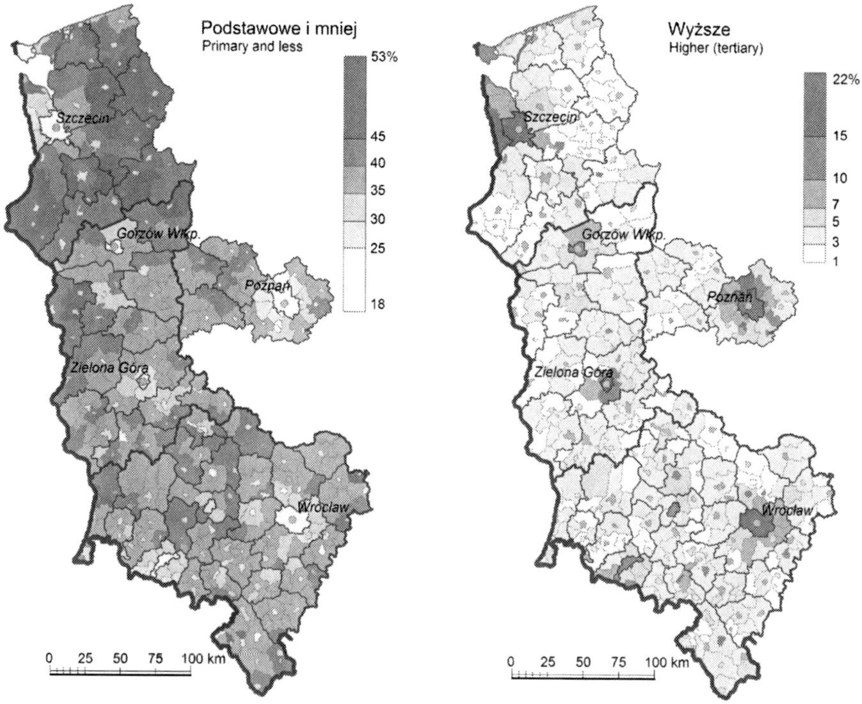

Fig. 6: Social development.

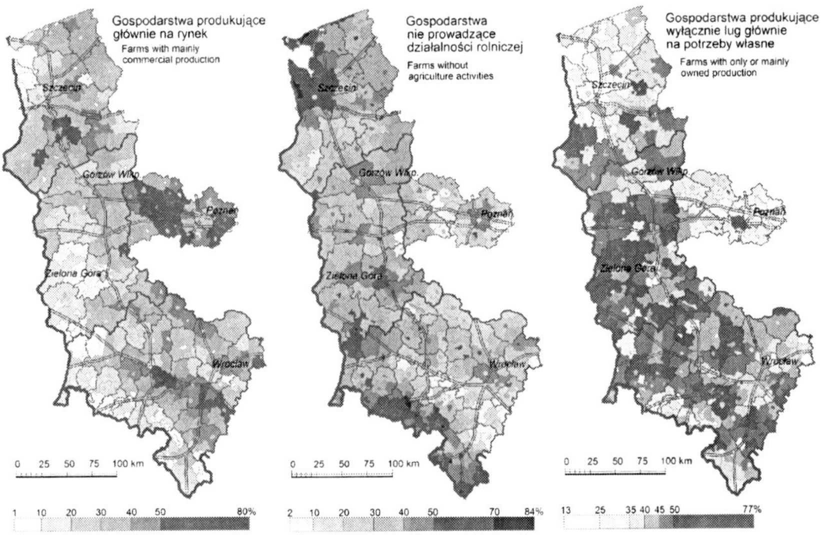

Fig. 7: Rural space and agriculture.

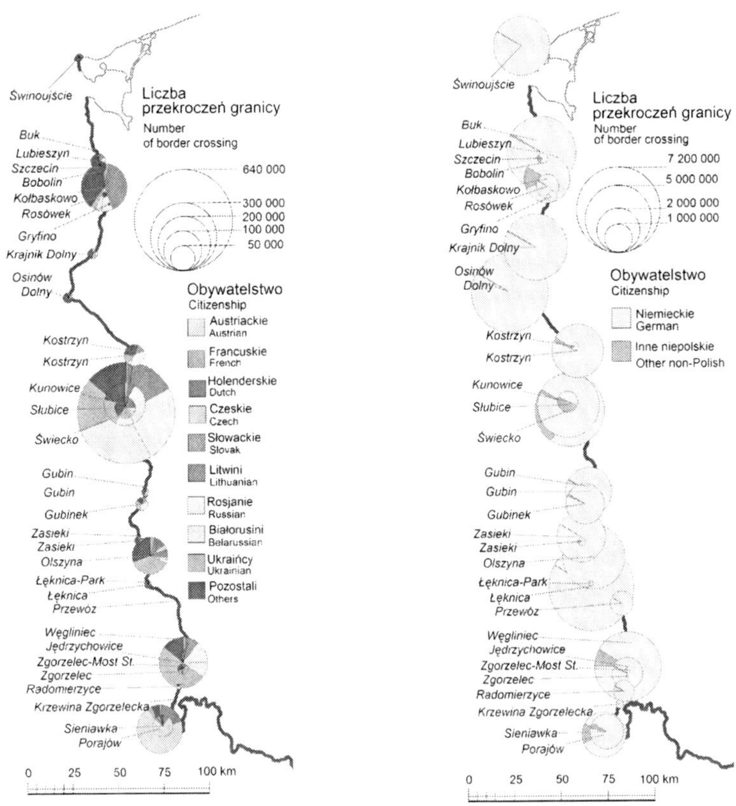

Fig. 8: Functional linkage on the national and transboundary scales.

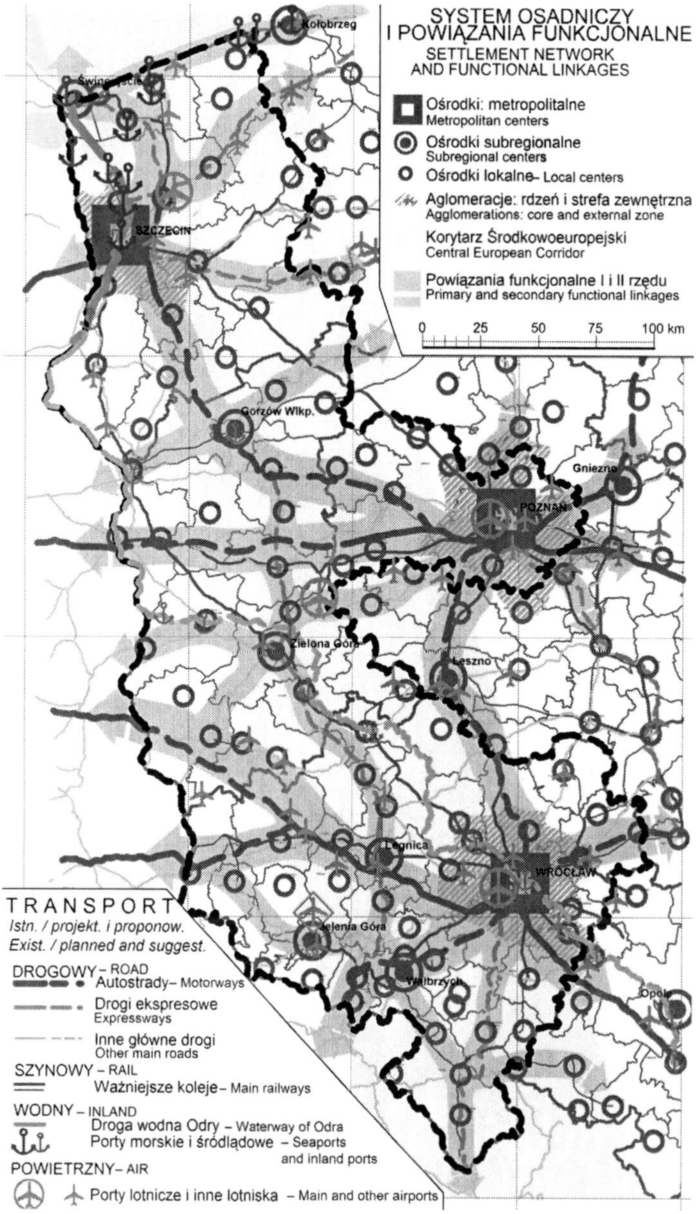

Fig. 9: Transport and technical infrastructure.

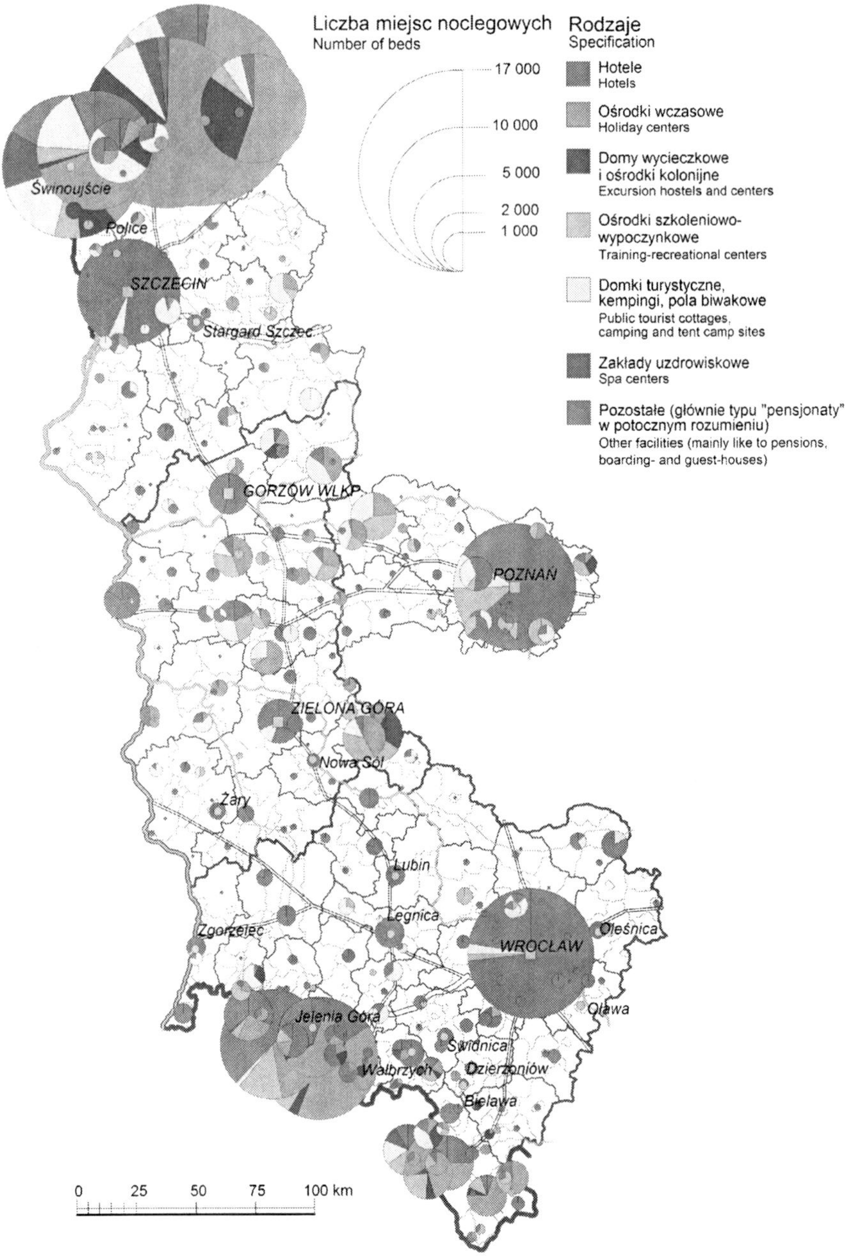

Fig. 10: Tourist infrastructure – centres and development.

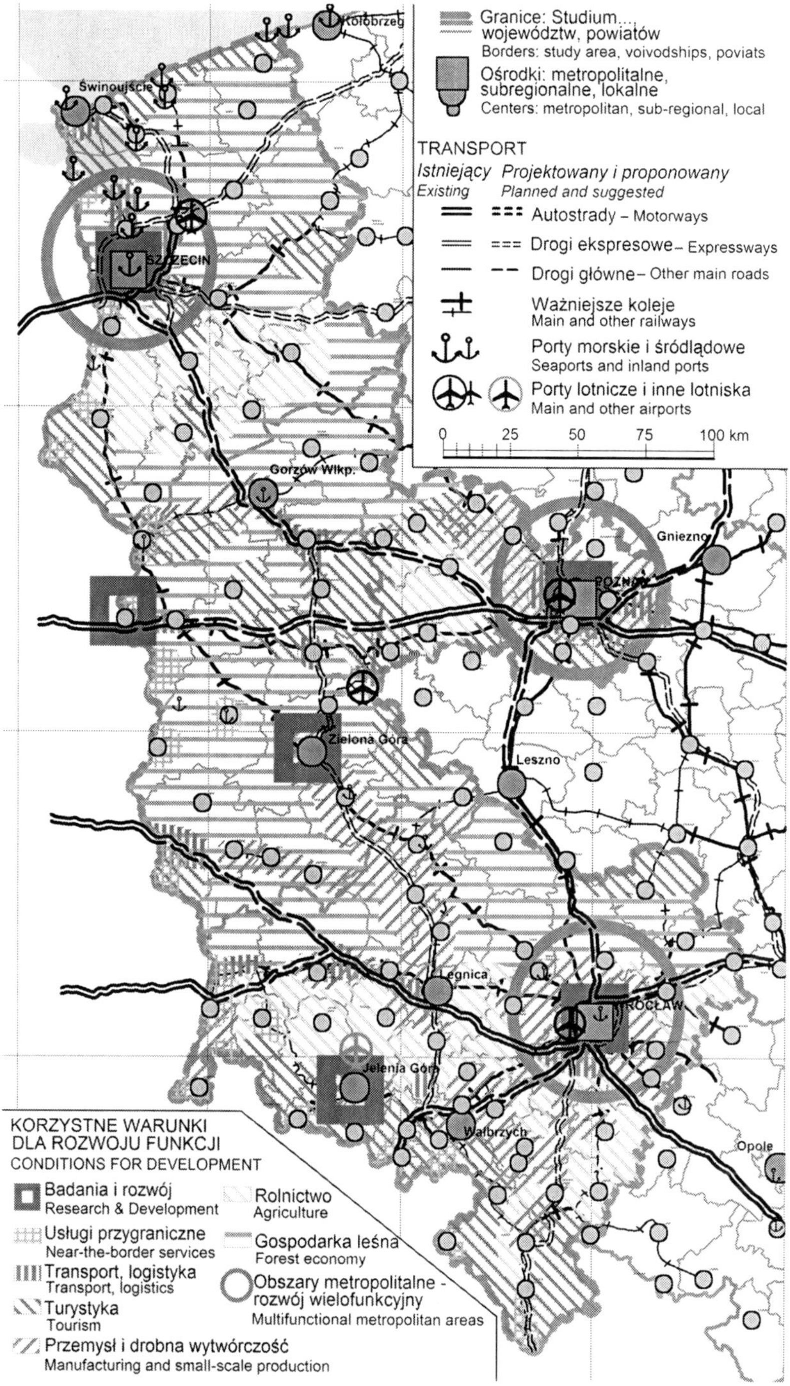

Fig. 11: Regional development – conditions for development.

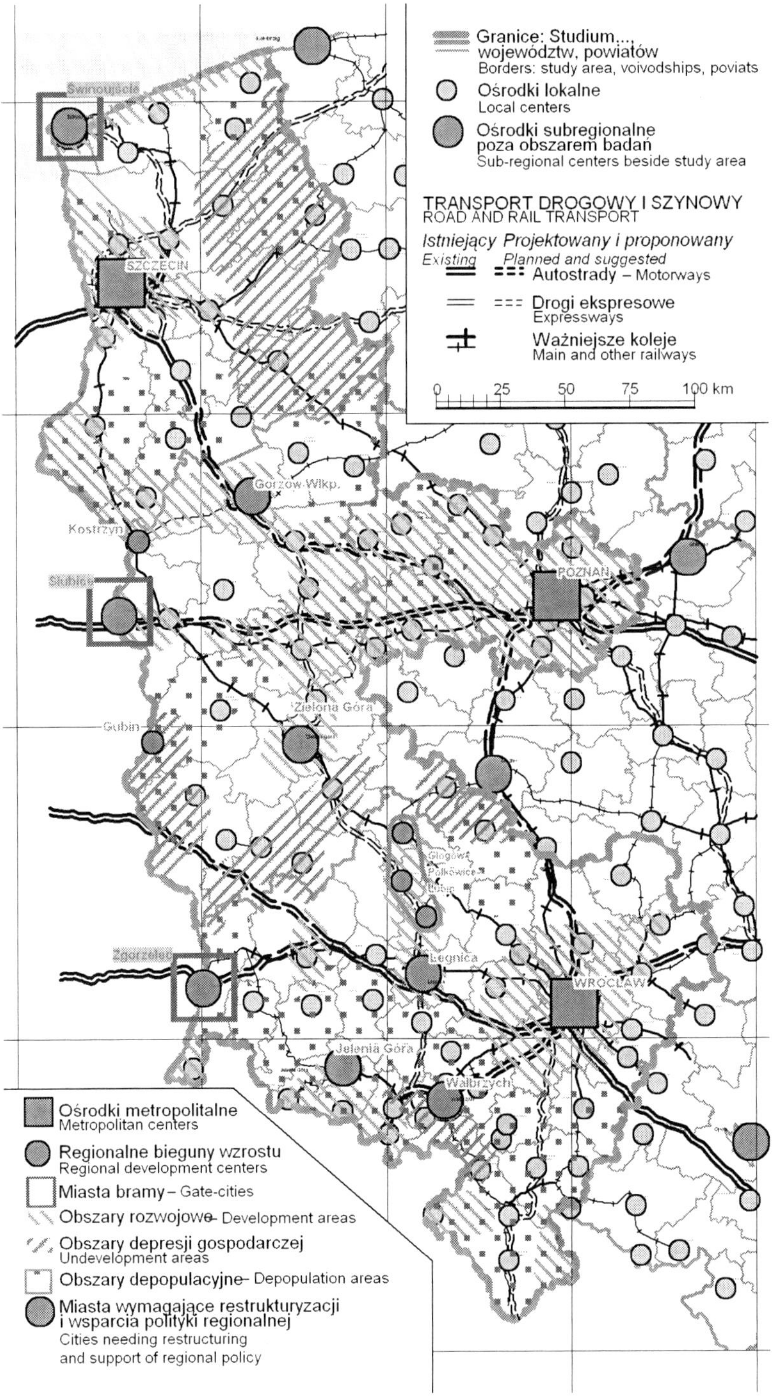

Fig 12: Regional development.

10.4 Direction of Future Development and Spatial Organization of the Borderland Area

The future of the western borderland area will be determined by the country's regional policy to a large extent. *The National Strategy for Regional Development 2007-2013* is particularly beneficial to the borderland region. As a matter of fact, all the strategic and operational objectives, as well as priorities, facilitate overcoming the disadvantageous peripheral status of the whole area addressed in the work - in order that it could play a leading role in United Europe in the future. It appears that, parallel to sixteen self-governmental Operational Programmes, an integrated operational programme covering three borderland voivodships along with Wielkopolskie voivodship ought to be framed, as suggested in regard to the operational programme of borderland and trans-boundary cooperation contained in the National Plan of Development 2003-2013.

Regardless of planning, investment processes must be implemented that would maximize the benefits from switching away from passive to active factors of endogenous economic development. This primarily requires concentrating on organizational and economic measures. In the near future, there exists a unique opportunity for significant growth in infrastructure investment. Correspondingly, financial outlays on education, science, research and development, as well as culture, should be significantly increased.

Education development and raising the level of science literacy is also a prerequisite for improvement in functional efficiency of regional self-government institutions, for growth in business services, and enlargement of regional policy institutions. At the same time, it constitutes the basis for exchange of international experiences, as well as basis for transfer of practices and skills. It ought to be stressed that there is a growing awareness for the necessity to invest in infrastructure and education, reflected in all planning and strategic documents. It is also clearly stated in *The National Strategy for Regional Development 2007-2013.*

Notwithstanding a great leap forward to improve the Polish-German cooperation after 1989 in terms of quality, there exists a necessity to take further radical measures on the part of local authorities, arising from a tangible threat of the borderland region becoming simply a transit area of inferior economic significance. There is also a danger of not taking advantage of the opportunities that emerged after the EU enlargement, as well as stirring up economic, cultural, and political conflicts within the borderland region.

The key issue in determining directions for development and spatial organization is to identify the principal actors interested in development of the area, along with working towards its main goals and interests. The central problem for the Polish economic entities constitutes severe restrictions being in force, hampering access to the German market of services. It is worth adding that in the post-World War Two period the chief group of interests were the Polish and Russian armies. The present-day identification of the main actors and awareness of their contradictory objectives is the first step towards arriving at a compromise.

The main centre for diffusion of innovation and economic integration on the Polish side within the Polish-German borderland region should be the metropolitan structure comprising Szczecin-Poznań-Wrocław. One of the targets of the country's regional policy should be strengthening infrastructural linkages between those metropolitan areas, allowing for free development of economic, social, and cultural relations. They may be competitive in terms of presenting themselves as viable alternative open to offers coming from the Berlin metropolitan area.

German plans dealing with the organization of the Polish-German borderland region cannot always be viewed as beneficial for the Polish side. This concerns in particular plans prepared in years 1990-1992: *Europeanization of Odra-Nysa Region* (author: Hans Koschyk), *Polish-German borderland cooperation. Thoughts and Conclusions* (author: Victor von Malthus), *Maquilador Programme* (the Friedrich Bert Foundation and American Institute for Studies on the Northern Borderland), *Conception supporting the Odra region* (Willers' Plan).

Establishing infrastructural links between Szczecin, Poznań and Wrocław, together with economic development of those metropolitan areas, may contribute to the creation of the axis of the Polish "boomerang of accelerated development" in Central Europe – as an alternative or the Polish version to the proposed Central European Boomerang between Prague, Dresden, Berlin and Szczecin. This conception is in keeping with the spatial policy of the EU stated in the ESDP, especially with such principles as: polycentric spatial development through the growth in significance of the junction or interchange elements of the settlement structure; and also through the increased importance of the infrastructure belts linking the European metropolises. The priority is, then, supplementing the target network of motorways and expressways by additional planning and construction of the motorway connections Szczecin-Poznań and Poznań-Wrocław. Investments in the projected expressway of transit significance alongside the western border of Poland seem to be of marginal importance in this context.

Diversified population dynamics form a decisive element in determining directions and changes in spatial organization. There are unquestionably greater population dynamics and demographic pressure on the Polish side, e.g. worse housing conditions coupled with dwellings shortages, leading to tensions in the light of large numbers of empty dwellings on the German side. A similar situation involves the larger business activity dynamics on the Polish side, faced with some tough restrictions affecting the activity of Polish business entities in the immediate vicinity of the German borderland. The huge differences between demographic potentials combined with differences in regard to labour costs and level of unemployment on both sides of the border may possibly generate considerable social and economic tensions in the future. In addition, economic potential discrepancies influencing the local authorities' capabilities for financing infrastructural investments place the Polish side in an unfavourable situation.

Tensions may also stem from the fact that the eastern areas of the German borderland will benefit less from EU enlargement than the western parts of Germany, while the Polish areas of the western borderland appear to be one of the major beneficiaries of the EU accession.

The zone immediately adjacent to the border, heterogeneous in nature, is still benefiting from trans-boundary trade. Thus, it constitutes the area where the actual level of unemployment is low. In order to maintain the high rate of growth, the economic entities operating within this area must invest considerable parts of their profits in creating a modern services infrastructure and in expanding the range of services offered, as well as undertaking up-to-date production ventures.

Heterogeneity of the area is reflected in several typologies and regionalizations contained in the plans. At present, within the area of the *study,* we distinguish metropolitan areas and medium-sized towns forming the regional growth poles as well as larger towns or cities that, in particular, require restructuring. As far as developmental areas are concerned, regions exist that are dominated by production functions (agriculture, small-scale production or tourist function). The majority of those areas are characterized by depopulation tendencies that are a reflection of the serious economic problems existing here. There are also areas affected by economic depression related to the collapse of the state-owned farms, as well as the collapse and restructuring of the industry.

The Position of the Polish-German Borderland within the Central European Regional Structure

The position of the Polish-German borderland based on the continental socio-economic structure can be characterized with a few distinctive features. This area is/has:

- centrally located at the continental scale,
- burdened with the problems remaining from previous epochs, especially from the World War II period and the communist era of centrally-planned economy, as well as by the sad memory of East German proximity,
- located on both sides of the typical subsequent border (Harsthorne 1936), separating communities marked by cultural differences, with inhabitants staying there for a relatively short period of time,
- performing the function of the external borderland in the period 1990-2004 and since 2004 the function of the internal EU borderland,
- situated in the area of influence exerted by the capital city of Berlin (however, this influence is of lesser significance than was originally assumed),
- partly belonging to Baltic Europe, thus it is related to maritime economy, constituting a supplying zone for the port basin Szczecin-Świnoujście,
- relatively good transportation accessibility from the economic heart of the EU, thus having superior access to the market for goods,
- relatively inferior accessibility (except for the central part) from the central part of Poland, primarily from Warsaw,
- determining the limits of labour costs within the EU,

- cut through by transit corridors going in directions such as E-W as well as NE-SW,
- specific cultural diversity of the population as a result of the fact that these regions – the so-called Regained Territories (incorporated into Poland after WWII) - were populated by migrants from the East,
- a major beneficiary of EU funds already in the pre-accession period,
- concentrating its international cooperation on the self-government level (four Euroregions: Nysa, Sprewa-Nysa-Bóbr, ProEuropa Viadrina, Pomerania),
- strong socio-economic relations with Germany (however the majority of those interactions involves rather the western part of Germany and only to a lesser degree the eastern part).

The importance of the above-mentioned elements is subject to dynamic changes. Along with the changes in border function as well as Poland's accession into the European Union, determinants of historic nature decreased in importance. Some minor economic relations within the borderland region had certain significance, which helped in eliminating psychological barriers. For the last 20 years, the eastern German lands have witnessed massive investments in infrastructure. This led to improvement of accessibility of the Polish side of the borderland and exerted stronger pressure for development transport networks on the eastern side of the Odra River.

Transcontinental transit movement as a factor stimulating the region's development proved less significant than it was originally assumed. Priorities of infrastructure development within this area ought to be identified in accordance with demand for internal and bilateral Polish-German transport. Transport involving third parties must be perceived as an important source of profits at the local scale (e.g. alongside the transport corridors), but it cannot be regarded as a major stimulus for development.

The main direction of transit flow is – and in all likelihood will remain– the route from the Baltic countries and Eastern Europe through Warsaw towards Berlin. Progress in construction of the A4 motorway only partly contributed to transferring the transport flows from the Ukraine to the southern corridor. North-east transport corridors are subject to more intense competition on account of the much better developed transport infrastructure existing in Germany, the Czech Republic, and Scandinavia (Danish bridges) than it is in the case of the East-West routes. Therefore, these investments must be planned with taking domestic or possibly German demand into consideration.

The Position of the Borderland within the Regional Structure of Poland

Throughout the last decades the area of the western borderland of Poland constituted a peripheral region situated in the neighbourhood of the well-guarded state border. It was also a spatial barrier - effective enough to blockade the exchange of people, services or capitals between Poland and neighbourhood countries.

In Europe today, processes of metropolization have undergone significant acceleration. The function of large developmental centres for higher-class services, as well as hierarchic diffusion of innovations, has become a factor determining development of adjacent areas. The modern and polycentric (hexagonal) transport and telecommunication network is a prerequisite condition for that diffusion. For that reason, the strengthening of linkages within the entire western borderland area, as well as between regional centres and the national metropolises, must receive priority. Only after achieving this goal, strong linkages with German centres should be established. The research areas comprise the following centres: Wrocław, Poznań, Szczecin, as well as Berlin and Dresden; however, in broader perspective also Prague, Warszawa (key element is establishing satisfactory connection with Wrocław) and Gdańsk (Baltic system). Construction of a modern road infrastructure (motorways, expressways) connecting above mentioned metropolises should not be delayed until after 2013.

Priority given for construction of connections between metropolises cannot be perceived as sign of attaching less importance either to the southern linkage related to the S-3 expressway or to the river Odra water route, or to the parallel railway tracks. This network of linkages, defined as the Central European Transport Corridor, must be carried out. Within the spatial context and from the perspective of *The National Strategy for Regional Development 2007-2013*, the idea of the Central European Transport Corridor ought to be perceived in much broader terms. This is in accordance with the priority of construction of supra-voivodship extensive functional-spatial and economic structures included into the aforementioned NSRD project. The essential significance includes establishing the supra-voivodship structures alongside the Baltic Sea, including intensified efforts to accomplish construction of the S-6 expressway Szczecin-Gdańsk.

Similarly, improvement in local infrastructure of the narrowly understood borderland belt must be of top priority. Also, of key importance is ensuring that areas situated between border crossings (gateway towns) have access to these crossings (for example, roads such as: Pieńsk-Łęknica, Świecko-Gubin with a crossing on the Odra river, the western ring road of Szczecin with a bridge in Police, tunnel crossing in Świnoujście). It is immensely important to adequately "tie together" the motorways and expressways now under construction with the regional transport network (with taking into account a sufficient number of interchanges or junctions). Particular attention must be given to the local tourist-transport infrastructure (e.g. bicycle routes, bridges for pedestrians, ferries, tourist paths).

There seems to be a sufficient number of border crossings on the Polish-German border; however, except for a few cases, such as the bridge in Urad and road points such as Dobieszczyn and Świnoujście-Garz. The function of gateway towns must be ascribed to the selected centres on the most important routes (Zgorzelec, Słubice, Szczecin, Świnoujście), whereas the remaining centres ought to be rather protected against the massive transit movement, focusing only on their own accessibility with respect to the Polish and European transport network.

It should be stressed that the minimization of trans-boundary road transport is worth striving for, and the primary objective must be to organize intermodal transport of goods and related with that terminals and logistic centres. Within this context, it is important to improve trans-boundary railway infrastructure (e.g. modernization of the Węgliniec-Horka section of the railway line, introduction of the two-system electric locomotives into the Polish-German main railway lines). Also, it is necessary to alter railway tariffs policy, especially with regard to passenger rail links between Poland and Germany.

The development of regional airports in Wrocław, Poznań, Szczecin, as well as in Babimost and Jelenia Góra (a potential new airport) into international complexes deserves considerable assistance.

The prospect of Poland joining the Schengen Agreement determines the potential spatial effects of changes in the magnitude of movement on the border inside of the EU. Joining the agreement will result in closing down all border checkpoints, as well as allowing for the possibility of crossing the border in any place. When taking this into consideration, it is inadvisable to carry on with modernization of the existing complexes. The increase in movement may require further development of trans-boundary transport infrastructure (also for pedestrians and cyclists). The cessation of passport control will have a specific effect on the labour market, parallel to those generated by the removal of customs clearance in 2004. This may lead to growth in economic activity in the areas immediately adjacent to the border (as a result of disappearance of restrictions and also psychological barriers). The very border crossings may potentially be utilized for other purposes (the same happened in Western Europe), for example, by turning them into service complexes for drivers or as petrol stations, restaurants, retail outlets trading with regional products, or as tourist information centres.

Recommendations

Assuming, in line with the National Strategy for 2007-2013, that a key element in the country's regional policy is support for endogenous development in the voivodships, it is possible to identify conditions regarding the aforementioned spheres of economic and social activity that can be deemed favourable for the borderland area. The proposed bases that underpin endogenous development would also seem to meet the criterion for long-term development postulated in the National Strategy.

It is the multifunctional metropolitan areas of Poznań, Wrocław, and Szczecin that receive the greatest importance. However, the full use of those metropolitan functions in the name of development in the borderland as a whole is first and foremost dependent on having an effective system of communications and transport on both the regional and supra-regional scales. A second thrust to action under the country's regional policy, in the name of stimulated endogenous development of the borderland area, entails the necessary stimulation of science and education as the elements most likely to support research and development. Apart from the metropolises, R&D may become the most important function behind the endogenous development of such centres as Zielona Góra, Słubice (together with Frankfurt/Oder) and Jelenia Góra. The sizes, geographical locations, and easily accessible recreational and tourist attractiveness of these centres all suggest opportunities from a form of development that attaches primary importance to the university function. Gorzów Wielkopolski and Legnica also have chances for evolving in regard to research and development, at least if they are able to attract high-tech industries.

The servicing of border traffic as a primary function in endogenous development concerns relatively small areas isolated from each other, widely dispersed along the border, and in the immediate vicinity of the crossings thereof. Transport and logistics form another specific element of endogenous development, but are again concentrated in the relatively small areas of transport/communications nodes. The development of industry is very much dependent on the speed with which the traditional sectors can be restructured and their output brought up to date. The traditional industrial centres continue to possess significant amounts of untapped development potential. Polish regional policy (e.g. as effected by a decisive investment impulse) could still restore many depressed industrial areas to the path of economic development. It is probable that the main role in any such revival would be played by small-scale manufacturing (to the extent that this would gain support from the instruments applied in the country's regional policy).

The prevailing functions are forest, agricultural, and tourist management. They are broad and of course associated with valuable features of the natural environment to a very significant degree. However, proper use of these features while ensuring implementation of sustainable development principles requires coordination of regional policy at a level above that of the gmina and the voivodship.

The most important element stimulating economic development should be an established transport system of a cohesive nature that allows for the development of functional linkages on different levels. The presence of modern transport and communications infrastructure needs to be thought of as the very foundation of any growth in the region's competitiveness, and as a precondition if socioeconomic integration with the country as a whole is to be achieved. On the other hand, the lack of certain key investments may put in place a barrier of fundamental importance that will tend to cancel out the positive effects otherwise accruing from integration into European Union space. In this situation, it is of key importance (in line with the European Spatial Development Perspective) that there be simultaneous and integrated development of transport systems at different hierarchical levels.

Irrespective of the directions and specialised thrusts that any future economic development of Poland's borderland may embark upon, particular attention needs to be paid to the development of higher-order services in the metropolitan areas, in the smaller centres, and in certain of the centres by the border. The major aim here will be to curtail the currently ongoing sapping of capacity on the Polish side that now benefits German travel agencies and air carriers. At present, a problem is posed, not merely by the fact that services on the German side are generally of superior quality, but also by the fact that spatial access to Polish providers of services in Polish towns and metropolitan areas is limited.

A fundamental role is to be assigned to the appropriate support and promotion of the Szczecin metropolitan area and city/port of Świnoujście as port centres servicing the Polish-German borderland and its hinterland (including the metropolises of Berlin and Poznań). There would seem to be no justification for those concepts that assume a transfer of certain transport functions of the Szczecin-Świnoujście port complex to smaller Baltic centres and inland ports on the Oder (Schwedt). While the future development of Szczecin's metropolitan functions cannot be seen to lie solely with transport functions, these can provide a basis for the growth of higher-order services (including specialist education and training) by themselves, as well as technologically advanced production linked, for example, with the maritime economy.

It is in the interests of the borderland area that developmental priorities and environmental functions be balanced. It is important for strategic-level spatial planning that the Polish part should not have its economic development restricted excessively by virtue of overly stringent environmental standards (such as those associated with the *Natura 2000* programme). The break in development of many years' duration cannot now represent a pretext for the enforcement of a development policy that is of equal stringency to the one applied in Western European countries, which previously took opportunities for development of their infrastructure (notably in transport) available to them. Any such policy would entrench the economic marginalisation of certain areas, ultimately not favouring the attainment of environmental objectives either.

It is also important that use within the system of socioeconomic development of near-border regions be made of their ecologically valuable features, and most especially the natural wealth of the Nysa-Oder corridor, which is unique on the European scale. This requires that local communities be encouraged to give their acceptance to the protection of the most naturally valuable areas, while at the same time necessitating a verification of environmental protection programmes, to ensure that these curtail the region's potential economic development to the most limited degree possible. One of the pathways to the attainment of such objectives entails the most precious natural features being treated as added value, as far as the socioeconomic balance of regional development is concerned, such that the rational way of utilising them economically is to leave them to nature. Rational utilisation should be understood as activity of a kind that allows for the preservation and protection of valuable environmental features, while at the same time allowing for development activity. Instruments of particular significance in ensuring that such activity goes ahead are environmental impact assessments for particular developments, and – on the basis of them – searches for the optimal solutions as regards locations.

Water management and the state of water purity pose problems for the Polish-German borderland. This is true of management problems such as those concerning flood protection and the utilization of the inadequate water resources within the

Nysa Łużycka basin, and that of the Upper and Middle Oder, and of matters relating to the poor sanitary state of surface waters. This constitutes a serious barrier to utilization and to regional development, thanks to hindered development of tourist, recreational, agricultural and settlement functions.

Monitoring and knowledge of the socioeconomic processes ongoing within each region and country should today be regarded as standard requirements for most better developed countries in the world. Particular attention is usually paid to regions in which there are specific features associated with the large scale of ongoing changes, negative features, or particularly rapid development. From the time of onset of the transformation the border areas of Poland have naturally been subject to particularly dynamic and multidimensional processes that have to be assessed in both positive and negative terms. It is this ambiguity of assessment that makes careful observation of ongoing phenomena so very important.

10.5 Conclusions

In summary, it should be underlined that the Guideline study fulfilled its primary purpose of serving as a handy tool for diagnosis as well as a useful basis for further planning actions. Opportunities and threats to the Polish part of the borderland area were identified. In addition, basic directions for development were framed, both at regional and local level. Still, the actual worth of the prepared document proved to be influenced by the external institutional factors. Among the major restrictions, the following issues should be mentioned:

- there is no clarity about legislation validating the Guideline study within the Polish system of spatial planning (related to frequent changes in legislation);
- delimitation of the area covered in the study is unsatisfactory and primarily politically oriented;
- there is insufficient coordination at the intergovernmental level that could ensure implementation of the study's recommendations within the borderland area simultaneously on the Polish and German side;
- the clash with the interests of lobby groups related with the western borderland (expressed, for example, in the press articles);
- emergence of conflicts of interests between gminas belonging to borderland belt and regions (voivodships), which led to contradictory conclusions derived from consultations carried out by the team at the different levels of administration;
- ambiguity of the mutual relations between the study and the points included into the voivodships' development strategy (questions arise whether it is possible to critically assess the latter – the team sometimes gives recommendations contradictory to the strategy's points).

Solving the aforementioned issues seems to be a prerequisite condition so that the prepared Guideline study (and also other similar documents that will be created in the future) could stand a chance of becoming a binding document actually framing the spatial development of the areas covered by research.

Analyses carried out to date have made it possible to draw attention to several problems that pose a potential threat, and thus require rapid and far-reaching investigation and analysis.

A matter of primary significance is the delimitation of the border area, on the basis of the real range of impact and scope of certain economic and social linkages that are identifiably distinct in comparison with the rest of the country. Work of this kind should provide a definitive answer to any questions regarding the very existence of a genuine borderland region. In relation to the state's regional policy objectives, the borderland should be delimited as a region in which specific forms of action need to be taken. In this case, there is a need to identify the degree of spatial aggregation of analyzed information (gmina, poviat, or sub-regional levels, etc.).

In these circumstances, the country's regional policy is itself an important subject for study, especially when it comes to the consequences for integration or disintegration of the country's territorial integrity. Regional plans and strategies (from the different administrative tiers) – and most especially those for borderland areas – should be assessed from the point of view of Poland's national interests and the interests of entities active on Polish territory.

A further research topic of strategic significance for the western borderland concerns the observation of developments as regards Polish-German economic and social links, and most especially those in which the Berlin metropolitan area exerts an influence on areas within Poland. The proximity of Berlin – and the huge economic and social potential that it entails – has much to offer, though it also gives rise to a number of threats. There is thus an absolute need for research on the potential development of the Szczecin, Poznań and Wrocław metropolitan areas, with a view to these coming to represent metropolitan centres of significance for western Poland's regional development through their being, in some sense, competitive with (an alternative to) Berlin.

Analyses should also assess opportunities for non-agricultural economic activity in rural areas. Earlier single-function development of these left a particularly indelible mark upon those communities in which the nationalised sector (of the State Farms) was dominant until recently. The extent and severity of rural poverty and social exclusion (in particular following the abrupt closure of the State Farms) demands rapid solutions. It is for this reason that the search for alternative sources of income represents another important study objective.

Cross-border trade and attendant passenger traffic need to be made subject to ongoing study, as does tourist traffic and its spatial distribution. It is of particular importance that the changing structure of expenditure on the part of Germans and other non-Polish nationals is fully appreciated. This would in turn allow for the identification of the service sectors that are developing.

The incessant emphasizing – in practically all planning and strategic studies – of the role of tourism and the (overly-optimistic) possibilities for development it brings with it, make it essential to conduct work on the sector's real significance for today's economy. Agrotourism needs especially careful study, since this sector is indeed developing dynamically, while the data describing it are incomplete and subject to large errors (of possibly more than 30%).

Water is an issue in the Polish-German borderland. This is true, *i.e.*, of such management-related matters such as flood protection and the utilization of what are insufficient water resources within the drainage basin of the Nysa Łużycka.

Environmental protection programmes need to be put into effect in such a way as to minimize limitations to potential economic development, while proposals for bringing valuable environmental features within development could be drawn up. Warning analyses regarding the environmental consequences of particular types of construction are of fundamental significance.

There is also a need for work on the consequences of the planned construction or ongoing reconstruction and remodelling of transport and communications infrastructure when it comes to changes in regional structure.

References

Ciok S. (2004): Pogranicze polsko-niemieckie. Problemy współpracy transgranicznej, Wrocław: Wydawnictwo Uniwersytetu Wrocławskiego.

Degórski M. (ed., 2006): Natural and Human Development of Poland – a geographical overview. Warszawa: IGIPZ PAN, PTG, UW, UJ, UŚ.

Gorzelak G., Bachtler J., Kasprzyk M. (eds., 2004): Współpraca transgraniczna Unii Europejskiej: doświadczenia polsko-niemieckie. Warszawa: Centrum Europejskich Studiów Regionalnych i Lokalnych UW, Wyd. Nauk. „Scholar".

Komornicki T. (1999): Granice Polski. Analiza zmian przenikalności w latach 1990–1996, Geopolitical Studiem, IGiPZ PAN, 5, 348.

Komornicki T. (2004): Impact of border regime institutions on transport network development In Central and Eastern Europe. In: Rietvald P., Stough R. R. (eds.), Barriers to Sustainable Transport. Institutions, Regulation and Sustainability. London, New York: Spon Press, 183-196.

Komornicki T. (2001): Changes in the role and permeability of Polish borders. In: Geographia Polonica Vol.74, No.1, IGiPZ PAN, 77-100.

Liszewski S. (ed., 2003): Możliwości i kierunki rozwoju turystyki w Dolinie Odry: Łódzkie Towarzystwo Naukowe: Katedra Geografii Miast i Turyzmu Uniwersytetu Łódzkiego.

Program for the Oder 2006 (2000): Strategy for Modernization of the Oder Water System. Wrocław.

Rössner T., Anisimowa G., Komornicki T., Miros K., Röttger A. (1998): Die Miteleuropäische West-Ost-Achse Sachsen-Schlesien-Galizien, Daten Fakten, Literatur Zur Geographie Europas 5/1998. Leipzig: Institut fur Länderkunde.

Stasiak A. (ed.) (n.f.i.): Podstawy rozwoju zachodnich i wschodnich obszarów przygranicznych Polski, Seria – 12 z./Biuletyn – Instytut Geografii i Przestrzennego Zagospodarowania PAN.

Studium kierunkowe zagospodarowania przestrzennego obszaru wzdłuż granicy polsko-niemieckiej (1994): cz. 1, 2, 3 (synteza) Analiza sytuacji wyjściowej IGPiK. Warszawa, Essen.

Węcławowicz G., Degórski M., Komornicki T., Korzeń Ja., Bański J., Korzeń Ju., Soja R., Śleszyński P., Więckowski M. (2006): Studia nad przestrzennym zagospodarowaniem obszaru wzdłuż granicy polsko-niemieckiej. Prace Geograficzne nr 207. Warszawa: IG i PZ PAN. (Studies on spatial development of the polish-german border region. Geographical Studies, No. 207)

Węcławowicz G. (1996): Contemporary Poland. Space and society, Changing Eastern Europe 4. London: UCL Press.

11 Concepts, Strategies and Guiding Principles for Europe's Territorial Development

Karl Peter Schön

"Territorial cohesion", in addition to economic and social cohesion, is the new goal of the European Union (EU), which in recent years has become an important orientation and guiding principle for spatial development and territorial integration in the EU. European integration and cohesion, as it is seen today, is thus much more than a mere matter of standardisation and harmonisation of formerly national provisions and regulations for the economy and society. European integration is rather increasingly seen as the very concrete integration of the European territory, as a process of European cities and regions growing together and deepening their relationships based on a multi-level governance approach.

In 2001, in its White Paper on European Governance[1], the European Commission demanded more adequate interaction in multi-level partnership and, with a reference to the European Spatial Development Perspective adopted in 1999, expressed the need to build a better partnership across various levels in three areas:

- Policy shaping: the Commission should ensure that regional and local knowledge and conditions are taken into account when developing EU policy proposals
- Flexibility: more flexibility in the means provided for implementing EU legislation and programmes with a strong territorial impact
- Coherence: the territorial impact of EU policies should be addressed and these policies should form part of a coherent whole

The widening of 'institutional' to 'territorial' integration eventually became apparent in the European Constitutional Treaty of 2004 and, after its rejection in the French and Dutch referenda, in the Lisbon Treaty of 2007, which, in addition to the objectives of economic and social cohesion, launched territorial cohesion as a new third objective of the European Union. Today, European spatial development policy is strongly based on territorial cohesion as its new central underlying paradigm.

As the word 'paradigm' already suggests, territorial cohesion is much more an orientation and a way of thinking than a fixed goal or concept. To associate territorial cohesion with singular clear-cut goals such as diminishing regional disparities would over-simplify and obscure the concept. 'Territorial cohesion' rather puts 'territory' on the agenda of European politics. And it spreads the message: "Territory matters!" Of course, in the light of territorial cohesion, any policy that has to deal with elements such as differences and disparities between territorial units (e.g. regions, cities, transnational areas etc.) also has to consider territorial characteristics

1 European Commission (2001)

and potentials in order to analyse territorial impacts of policies. But what adequate policies in this respect are is a matter of discourse and careful consideration rather than of a schematic goal system.

Territorial cohesion is often discussed in relation to concepts such as "equivalent living conditions" ("Gleichwertige Lebensbedingungen") and to subjects such as "services of general economic interest" ("Daseinsvorsorge") in the EU terminology. Indeed, the "services of general economic interest" were an important issue that provided for opening doors for integrating the concept of territorial cohesion in the Lisbon Treaty.[2] The role of the services of general economic interest in promoting social and territorial cohesion was already acknowledged in previous EU treaties.

And, as Kersten[3] points out, territorial cohesion in this spirit, i.e. with regard to ensuring basic services of general economic interest in all parts of the EU territory, including its periphery and giving citizens the chance to adequately participate in the social life of the Union, is a guiding principle that tolerates differences but demands basic access to chances. In this context, Kersten argues, the European guiding principle of territorial cohesion might be a better future option for a shrinking society like Germany than the ambitious goal of "Gleichwertige Lebensbedingungen".
The European policy of territorial cohesion is supposed to

- contribute to a more harmonious development of the EU,
- be informed by and based on scientific evidence,
- consider territorial impacts of EU policies at an early stage,
- take into account territorial specificities when developing and implementing EU policies,
- integrate territorial aspects into sectoral policies and foster their coherence,
- promote the cooperation of cities and regions beyond national borders,
- include a continuous dialogue with public and private stakeholders of territorial development.

Under the Lisbon Treaty, the policy of territorial cohesion would become a shared competence between the European Union and the EU member states. This would give the European Commission the legal authority and competence to be the driving force in this field. However, in accordance with the EU's subsidiarity principle, the Member States and their cities and regions have their say in this. Yet, before the Lisbon Treaty comes into effect, territorial cohesion is mainly subject to initiatives from the Member States and will be further practised in cooperation between the EU member states and the European Commission.

But independently from the further ratification process of the Lisbon Treaty, the EU ministers responsible for spatial planning and territorial development have already adopted 'territorial cohesion' as an important common objective in the Territorial Agenda of the European Union (TAEU). And they have affirmed their

2 J. Kersten (2008)
3 ibid.: 5

commitment to work together even more closely, among each other and with the EU institutions, in pursuing the goal of territorial cohesion – "independently of the ongoing discussion about the EU's reform process"[4].

In an expertise commissioned by the Federal Office for Building and Regional Planning (BBR)[5] Battis and Kersten pointed out that the willingness of the EU member states to accept and adopt the concept of territorial cohesion was of decisive political relevance also for the European Commission: Without this, the implementation of territorial cohesion in the European legislation would become questionable in the light of the failure of the European Constitutional Treaty in the French and Dutch referenda.[6]

In any case, with or without the ratification of the Lisbon Treaty, the policy of territorial cohesion needs

(a) a common identification of the main European territorial challenges and policy responses based on scientific knowledge and political debate among the territorial actors at different levels, and
(b) the development of new concepts of European territorial governance to tackle these challenges and to implement adequate policy strategies.

In the following, this article will focus on these two issues and explore them in detail. However, before this is done, there will be a short review of the genesis of a European spatial development policy, which by now can be traced back for almost two decades.

11.1 A Short Review of the Genesis of a European Spatial Development Policy

Almost 20 years ago, in Nantes 1989, the ministers responsible for spatial development in the EU member states met for the first time for an informal ministerial meeting and discussed topics relevant for European spatial development. In this meeting, they agreed upon a more intensive and continuous cooperation in order to take part more actively in shaping the spatial development of the EU territory. This emerging European spatial development policy was encouraged and supported by the European Commission, in particular by its president Jacques Delors, who was not only an enthusiastic supporter of further European integration, but also a protagonist of the "European Social Model" and of common visions and strategies for a more balanced European development.[7] And in the years to come, the European Commission and the EU member states advanced the new European spatial development policy. But it was not only cooperation that drove this new policy field; it was also political competition between the European Commission and the EU member states, partly

4 Territorial Agenda of the European Union (TAEU) (2007: Chapter I, no. 6)
5 Battis U., Kersten J. (2008a) cf. also: Battis U., Kersten J.: (2008b: 201-207)
6 ibid.: 203
7 Delors (2004)

based on the fear of some Member States, including Germany and its *Länder*, that the European Union might occupy fields of competencies that are not covered by the EU treaties.

The starting point of this process in 1989 can be located within the history of EU integration, the particular ambivalent and strained situation – as it often was – and still is – of the European Union between increased disparities (by "enlargement") and increased integration of the EU (by "deepening") at the end of the 1980s decade; i.e. the tension between enlargement (which brings about more disparities, more heterogeneity) and deepening (more integration, more common rules, depending on a certain degree of homogeneity). The enlargement of the 1980s by three southern countries, Greece (in 1981), Spain, and Portugal (both in 1986), besides the many other aspects brought along totally new and large disparities for the EU never seen before to that extent within the EU territory. Most of these southern accession states and, in particular, many of their regions had an economic output (GDP per capita) of less than 75 per cent of the EU average at that time. Nevertheless, and one might say, despite this enlargement round and the new regional economic disparities, the 1980s were also the time of the further deepening of Europe's economic integration and of the completion of the internal EU market. To make divergent things fit, new European policy fields were established or further developed: the EU's cohesion policy, the regional policy and the Structural Funds, the Trans-European Networks, environmental policies and other policies and instruments to support the territorial integration of Europe. All these new or extended policy fields and elements were largely fixed in two documents: the Single European Act (SEA) of 1987 and the Maastricht Treaty of 1993.

In response to this ambivalent situation of deepening economic and political integration vs. coping with large regional differences and disparities in the EU, which were considerably reinforced after the opening-up of Eastern Europe and extended through the eastern enlargements of 2004 and 2007, the ministers responsible for spatial planning and territorial development of the EU member states started their intensive cooperation process and jointly developed a European spatial development policy. A milestone of this cooperation was the adoption of the European Spatial Development Perspective (ESDP)[8] on 10/11 May 1999 in Potsdam.

In the ESDP, central political principles, objectives and strategies of the European spatial development policy were fixed. According to them, the European spatial development policy does not justify any new competences on the EU level, but takes the subsidiarity principle into account and is based on the voluntary cooperation and national responsibility of each Member State.

With the ESDP, the EU member states and the European Commission agreed upon joint spatial objectives and guiding principles for the future development of the European territory. A central general objective is the balanced and sustainable development of the EU territory based on a polycentric settlements development.

8 European Commission (1999)

The spatial development policy thus claims and intends to contribute to the main objectives of European policy: to economic and social cohesion, to the conservation and management of natural resources and cultural heritage, and to a more balanced competitiveness of the European territory.

Following the agreement upon the ESDP (Potsdam) and the action programme (Tampere) in 1999, cooperation in the field of spatial development policy has concentrated on the implementation of the ESDP. In the years after Potsdam/Tampere – and partly already before in the framework of pilot projects supporting the ESDP process – three actions in particular were successfully implemented:

- the inclusion of all European states, beyond EU membership, in the debate and the application of the ESDP on the larger European territory in the framework of the European Conference of Ministers responsible for Regional Planning (CEMAT) and the Council of Europe,
- the launch and perpetuation of an EU programme on transnational cooperation (INTERREG III B) and
- the establishment of a European network for spatial monitoring and policy advice (European Spatial Planning Observation Network - ESPON).

Already during the 12th Session of the CEMAT in 2000 in Hanover, the Guiding Principles for Sustainable Spatial Development of the European Continent („CEMAT Guiding Principles")[9] were adopted. They were approved at the beginning of 2002 by the Committee of Senior Officials of the Council of Europe, so that the main guiding principles of the ESDP on the EU territory also provided orientation for all member states of the Council of Europe.

With the INTERREG III B programme, a very successful implementation of the ESDP agreements was started at the level of transnational cooperation of cities and regions in larger European cooperation areas.[10]

Following the successful test stage of the Study Programme on European Spatial Planning (1998-2000)[11], the ESPON 2006 programme for establishing a network of research institutes for analysis and observation of European spatial development was started in 2002.

After Potsdam/Tampere (1999), especially three events were of great importance for European spatial development (without the spatial development ministers having a hand in these events):

- In 2000 the European Council adopted the Lisbon Strategy, which was followed by the Gothenburg Strategy one year later and revised in 2005. With these two strategies, new priorities were agreed upon for European development: to strengthen the competitiveness and innovation potential of Europe and to create more growth and employment based on the principle of sustainable development.
- The EU enlargements of 2004 and 2007 by twelve, mainly Central and Eastern

9 European Conference of Ministers responsible for Spatial Planning (CEMAT) (2000)

10 for up-to-date information see www.interreg.de

11 BBR (2001)

European countries caused important changes in the economic, social and spatial constellation of the European Union. Apart from economic and social disparities, the territorial and regional diversity in the EU as well has considerably increased due to the enlargement and has attracted the attention of the political sector and the public.

- Finally, in the EU Constitutional Treaty, the EU member states in 2004 agreed upon establishing territorial cohesion apart from economic and social cohesion as an equal third EU objective. After the rejection of the EU Constitutional Treaty in the French and Dutch referenda, territorial cohesion became part of the Lisbon Treaty of 2007.

TREATY ON THE FUNCTIONING OF THE EUROPEAN UNION (TFEU)[12]

Title XVIII
Economic, social and territorial cohesion

Art. 174 [ex 158] I In order to promote its overall harmonious development, the Union shall develop and pursue its actions leading to the strengthening of its economic, social and territorial cohesion.
II In particular, the Union shall aim at reducing disparities between the levels of development of the various regions and the backwardness of the least favoured regions.
III Among the regions concerned, particular attention shall be paid to rural areas, areas affected by industrial transition, and regions which suffer from severe and permanent natural or demographic handicaps such as the northernmost regions with very low population density and island, crossborder and mountain regions.
Art. 175 [ex 159] I 1 Member States shall conduct their economic policies and shall coordinate them in such a way as, in addition, to attain the objectives set out in Article 174. 2 The formulation and implementation of the Union's policies and actions and the implementation of the internal market shall take into account the objectives set out in Article 174 and shall contribute to their achievement.
3 The Union shall also support the achievement of these objectives by the action it takes through the Structural Funds (European Agricultural Guidance and Guarantee Fund, Guidance Section; European Social Fund; European Regional Development Fund), the European Investment Bank and the other existing Financial Instruments.
II 1 The Commission shall submit a report to the European Parliament, the Council, the Economic and Social Committee and the Committee of the Regions every three years on the progress made towards achieving economic, social and territorial cohesion and on the manner in which the various means provided for in this Article have contributed to it. 2 This report shall, if necessary, be accompanied by appropriate proposals.
III If specific actions prove necessary outside the Funds and without prejudice to the

12 Consolidated versions of the Treaty on European Union and the Treaty on the Functioning of the European Union (2008)

measures decided upon within the framework of the other Union policies, such actions may be adopted by the European Parliament and the Council acting in accordance with the ordinary legislative procedure and after consulting the Economic and Social Committee and the Committee of the Regions.
Art. 176 [ex 160] The European Regional Development Fund is intended to help to redress the main regional imbalances in the Union through participation in the development and structural adjustment of regions whose development is lagging behind and in the conversion of declining industrial regions.

These political trends in Europe in general and the new territorial cohesion objective in particular induced the Dutch EU Presidency in 2004 to take the new political initiative which aimed at adding more weight to the territorial dimension of European decisions. This was taken on by the Luxembourg EU Presidency. The ministerial meetings of Rotterdam (2004) and Luxembourg (2005) marked the beginning of a new understanding of spatial development policy with a stronger link to knowledge and governance (see the following chapters). The process eventually led to the adoption of the Territorial Agenda of the European Union (TAEU) and the Leipzig Charter on Sustainable European Cities (LC) in 2007. Both political documents were adopted at the Informal Ministerial Meeting on Urban Development and Territorial Cohesion held in Leipzig on 24-25 May 2007. Based on this, the First Action Programme for the Implementation of the Territorial Agenda of the European Union was adopted at the Informal Ministerial Meeting on Territorial Cohesion and Regional Policy held in Ponta Delgada (Azores) on 23-24 November 2007 (TAEU-AP).

11.2 Knowledge-based Identification of Territorial Challenges

The ESDP implementation, especially through the programmes ESPON and INTERREG, helped to gather new scientific findings on European spatial development and collect various experiences from the cooperation of cities and regions in international and transnational contexts. The ESDP had already shown elements of an evidence-based approach[13], even though, according to the knowledge

13 The Rotterdam ministerial meeting brought about the vision that the further development of a European spatial development policy should be based on existing scientific knowledge and evidence. The notion of 'evidence-based policy' has been used before, particularly by the Blair Government in the UK, and the idea that spatial development policies could be coordinated by the use of scientific information has also been a familiar idea in Germany for a long time ("Koordination durch Information"). However, the concept of an 'evidence-based policy' is by far not undisputed as it might suggest a positivistic perception of a supposedly rational and mechanistic policy (cf. Davoudi 2006) which is of course far from reality. Nevertheless, scientific discourses and practical experiences do play a role in policy formulation and, to explore this role in the context of European spatial development policy here, the broader term 'knowledge-based' is used.

basis on European spatial development at that time, the evidence basis was much smaller[14]. Since the adoption of the ESDP (1999), the knowledge about European spatial development and the experiences drawn from European cooperation projects have increased considerably. Both programmes delivered useful findings regarding the current Territorial Agenda[15]. Within the ESPON alone, 25,000 pages with text and hundreds of informative maps and illustrations on 35 thematic projects on spatial development in Europe were produced. In the INTERREG III B cooperation areas, more than ten thousand actors worked together in transnational projects. In the cooperation areas with German participation alone, 6,500 partners, 1,000 of them being German partners, cooperated in around 500 projects[16].

In Rotterdam and Luxembourg the ministers agreed upon using these experiences systematically and attaching more importance to new experiences and research results gained in recent years. The European spatial development policy was supposed to become "evidence-based", i.e. to be based on a new and deeper relationship between research and political strategic development. Stimulated by the European spatial development policy and the ESDP, a pool of knowledge and experiences was developed which again can flow back into the political development processes. Three expert documents were submitted as background documents on the Territorial Agenda to the Ministerial Meeting on 25 May 2007 in Leipzig: the document "The Territorial State and Perspectives of the European Union" elaborated by a European team of experts, as well as two BBR documents presented by the German EU Council Presidency: "Bringing Europe together" and "Maps on European territorial development"[17].

The "Territorial State and Perspectives" document provides an analysis of the initial situation and development perspectives mainly based on ESPON results and takes into consideration the experiences and innovative approaches of action gained from INTERREG projects in transnational cooperation. The paper takes on the issue of the new territorial challenges for Europe, such as climate change and energy supply and consumption, demographic change, including the depopulation of rural areas and other regions of the EU, growth and innovation, and documents them - from a regional perspective - with maps and project examples. It also takes on the discussion about the role of metropolitan regions and the uneven regional effects of globalisation including the widening gap between rich and less prosperous regions, as well as the unequal development of urban and rural areas combined with structural change. And it identifies new fields and necessities of action of spatial

14 cf. Böhme K., Schön P. (2006)

15 see contributions of Ahlke B., Görmar W., Hartz A. (2007) on transnational cooperation (INTERREG III B) and Schmidt-Seiwert V., Porsche L. (2007) to Maps on European territorial development

16 see Ahlke B., Görmar W., Hartz A. (2007)

17 sources and links to Internet downloads http://www.bbr.bund.de/cln_007/nn_169708/EN/ResearchConsultation/SpatialPlanning/SpatialDevelopmentEurope/EuropeanSpatial_20DevelopmentPolicy/01__Publications.html

development policy in areas such as risk management, urban sprawl and land use, infrastructural improvement, innovation, and cluster policies. The analyses are complemented by the "Maps on European territorial development" submitted by the German EU Council Presidency, which deal with the results in more detail and back them by other sources (e.g. Urban Audit results)[18].

> (7) We are facing major new territorial challenges today. These include:
> - regionally diverse impacts of climate change on the EU territory and its neighbours, particularly with regard to sustainable development,
> - rising energy prices, energy inefficiency and different territorial opportunities for new forms of energy supply,
> - accelerating integration of our regions, including crossborder areas, in global economic competition, and at the same time increasing dependencies of states and regions in the world,
> - impacts of EU enlargement on economic, social and territorial cohesion, particularly with regard to the transport and energy infrastructure related integration of Eastern Europe and the new EU Member States as well as their regions,
> - overexploitation of the ecological and cultural resources and loss of biodiversity, particularly through increasing development sprawl whilst remote areas are facing depopulation,
> - territorial effects of demographic change (especially ageing) as well as in and out migration and internal migration on labour markets, on the supply of public services of general interest as well as the housing market, the development of the settlement structure and how people live together in our cities and regions. (TAEU, Chapter I, no. 7)

Six objectives which have been agreed upon politically are set out in the Territorial Agenda: the establishment of a polycentric, balanced urban system; the creation of a new urban-rural relationship; the formation of innovative regional clusters; ensuring parity of access to infrastructure and knowledge; the promotion of trans-European risk management; sustainable development, forward-looking management and conservation of natural and cultural heritage.

It is also a new approach of the territorial development policy to seek for a close combination and cooperation with urban matters. The objectives set out in the Leipzig Charter on Sustainable European Cities make greater use of integrated urban development policy approaches by creating and ensuring high-quality public spaces, modernising infrastructure networks and improving energy efficiency, promoting proactive innovation and educational policies and – particularly in deprived neighbourhoods – promoting sustainable, efficient and affordable urban transport, pursuing strategies for upgrading the physical environment, strengthening

18 see Schmidt-Seiwert V., Porsche L. (2007)

the local economy and local labour market policy as well as proactive education and training policies for children and young people.

11.3 New Concepts of Territorial Governance

The Territorial Agenda of the EU wants to pave the way for a new understanding of development and planning. The objective "territorial cohesion" must not to be understood as one fixed objective, e.g. the reduction of regional disparities, but as a permanent process of political, organisational and technical cooperation between all actors of spatial development (TAEU, Chapter I, No. 4).

In the Agenda, the EU ministers refer to this cooperation process as "territorial governance": the economy (especially local and regional entrepreneurship), the scientific and administrative sector (mainly local and regional authorities), non-governmental organisations and sectoral policies have to cooperate in order to fruitfully use the urgently required investments in European regions (TAEU, Chapter I, No. 5).

Apart from the evidence-based planning method, the participatory character of the new European development and planning idea is tested for the first time in elaborating and discussing the Territorial Agenda. The keyword is "stakeholder dialogue". While the European Spatial Development Perspective (ESDP) was primarily developed by the Member States in close cooperation among themselves and with the European Commission, this time the Territorial Agenda drafts should be available to the public from the very beginning. A large continuous dialogue with the public should be initiated so that the Member States would be able to take the results of the stakeholder dialogue into account during their consultations.

The decisive factor for the success of this new European understanding of spatial development policy is the finding that territorial cohesion means much more than the development of new forms of interregional or interlocal cooperation as it is described in the ESDP ("a New Urban-Rural Relationship"). This horizontal cooperation of territorial authorities is inevitable in order to jointly formulate regional and subregional development strategies. The Territorial Agenda with its ideas goes beyond those of the ESDP when it emphasises that, in the context of a territorial cohesion policy, the cooperation of territorial authorities should be supported by a stakeholder dialogue with other actors of spatial development.

The German EU Council Presidency aimed at involving the European public in considerations concerning the Territorial Agenda from the very beginning. However, this could not be based on any experiences on how to organise this process. Federal Minister Wolfgang Tiefensee opened the Europe-wide debate during a stakeholder conference at the end of June 2006 in Amsterdam, following the invitation of his Dutch counterpart. The debate included an Internet-based dialogue and an exchange of opinions with European institutions, non-governmental organisations, and companies. It actively involved the European regional development actors in the working process.

The approximately 100 stakeholders of the Internet dialogue were composed of (European) non-governmental organisations, local and regional authorities, universities, foundations, local umbrella organisations and chambers from a variety of EU member states. The comments had an influence on the discussion and the final version of the TAEU document. For instance, there was a heated discussion about the role of small and medium-sized towns as important regional centres and, generally, about the role of rural areas in different parts of the EU. In the end, the TAEU put much more emphasis on the small and medium-sized towns than it had done in the first drafts provided for discussion and than can be found in other documents on territorial development.

Furthermore, there was a lively exchange with the European Commission and with the other EU institutions. The Committee on Regional Development of the European Parliament, the Committee of the Regions, the European Economic and Social Committee, the European Investment Bank, and the European Conference of Ministers responsible for Regional Planning (CEMAT) already actively participated in the dialogue and formulated comprehensive opinions before the Ministerial Meeting in Leipzig. Also, the Territorial Agenda was discussed during the annual conferences of a number of non-governmental organisations.

During two directors-general meetings and several workshops, the EU member states agreed upon integrating important results of this Europe-wide debate into the final draft of the Territorial Agenda of the EU. This especially included the validity of the ESDP objectives for the enlarged EU with 27 Member States. Furthermore, the objectives of the Lisbon and Gothenburg Strategies were accepted to be of equal political value for measures of spatial development policy. The directors-generals also pleaded for dovetailing the Leipzig Charter on Sustainable European Cities and the Territorial Agenda more strongly.

11.4 The Dialogue with Corporations as a Part of the Stake holder Process

With the stakeholder dialogue on the Territorial Agenda, new forms of involving the economy into the political process were tested as well. It helped to use the knowledge of leading corporations in the private economic sector for developing the document and better coordinating the interests of public and private actors.

The economic development is based on entrepreneurial decisions which can mainly be influenced by the public sector by creating attractive framework conditions. The competitiveness of a region increases with its ability to attract companies. An essential interest especially of local and regional authorities is to urge companies to pursue a sustainable investment policy. In the Territorial Agenda of the EU it is assumed that the more private stakeholders identify with the spatial development concepts of regional governments and administrations, the better they will manage. Spatial development policy faces special challenges here, as the territorial ties of corporations loosen increasingly in the age of globalisation. Cities and regions therefore face the question how relations to local companies or firms, in the attempt to win them over to a location, could be intensified. New forms of cooperation between regions and the business sector are therefore required.

Against this background, the dialogue with companies was sought while drafting the Territorial Agenda. It was difficult, as none of them participated in the Internet-based dialogue. The German Council Presidency therefore purposefully contacted companies. This time, the economy should increasingly be given opportunities to bring their experiences and concerns into the political process. A concept concerning the dialogue with stakeholders from the private industry has thus been developed.

In the framework of the research project "Main features of a territorial cohesion policy"[19], representatives from 30 companies participated in four workshops and discussed central topics and strategies of the Territorial Agenda:

- settlements development (representatives from retail trade and the housing industry)
- knowledge-based society (representatives from the cultural and media sector)
- mobility (representatives from transport and tourism companies)
- networking (representatives from telecommunication and energy companies)

Based on maps and analyses of the European spatial structure and an introduction given into alternative spatial development strategies, all four workshops included intensive, well-informed, and sometimes also controversial discussions, which very concretely referred to the draft Agenda. Many representatives from the business world emphasised that also on the Single European Market, which is growing together, the regional or local ties of their enterprises were of high importance. They expect the Territorial Agenda to give an incentive which includes involving companies in local and regional development planning at an early stage ("Round Table Local Economy").

As a follow-up activity, an introductory conference is planned for this year's Euregia in Leipzig (27-29 October 2008), the big communication forum on regional development in Europe, to deepen the strategic dialogue between political and economic decision-makers with regard to territorial development. The idea is to establish a common platform for a more continuous exchange and dialogue. Furthermore, this conference might give a starting signal to test economy and politics within concrete model projects in having a better share in a sound and sustainable territorial development.

11.5 Outlook

The process of agreeing upon common concepts, strategies, and guiding principles for Europe's territorial development is ongoing. The relevant actors on all levels

19 for further information on the research project please see: http://www.bbr.bund.de/cln_007/nn_62854/EN/ResearchProgrammes/GeneralDepartmentalResearch/SpatialPlaning/CohesionPolicy/01__Start.html

– at European, national, regional, and local levels – continue to shape this new 'multi-level system of territorial governance'. The public actors open up the dialogue to include not only non-governmental organisations, but also economic decision-makers and the civil society. With the support of the European Union, the new Structural Funds Objective 3 "European Territorial Cooperation" is being implemented with the research network ESPON 2013 and transnational cooperation (Interreg IV B) as two elements of particular importance for territorial cohesion and European spatial development policy. The former contributes more scientific knowledge to the sharp analysis of the great challenges of territorial development in Europe; the latter provides practical solutions and experiences to cope with those challenges.

The ESPON 2013 Programme will implement some new elements that are directly linked to the Territorial Agenda. Firstly, more emphasis shall be put on developing methods and models to analyse the territorial effects and impacts of EU sectoral policies. Secondly, new territorial indicators shall be developed to continuously monitor the EU's territorial development against the background of objectives such as territorial cohesion. This can be seen as one further step towards preparing for the application of the 'Open Method of Coordination' in territorial cohesion policies.[20] Thirdly, with the new 'client-oriented approach', cities and regions, authorities and stakeholders are more actively involved in the European research network.

With reference to the Lisbon Treaty and following the explicit wish of the EU member states expressed in the Leipzig meeting and the Territorial Agenda, the European Commission is currently preparing a Green Paper on Territorial Cohesion, which is supposed to be presented in autumn 2008. This Green Paper is expected to contribute to the discussion and support a common understanding on the main elements of the new objective "territorial cohesion" and to draw up suggestions on how to shape and implement a territorial cohesion policy.

The implementation of the Territorial Agenda and the Leipzig Charter is progressing. Several working groups have been established. They attempt to bring together urban and territorial questions and explore the crossing of sectoral and spatial policies.

And last, but not least: Themes of spatial development are becoming congruent in debates all over the world. Many crucial questions of today, economically (e.g. world-wide division of production), socially (e.g. world-wide competition of labour supply and demand), ecologically (e.g. global climate change, transport and energy demand) are becoming questions of worldwide organisation. They are also becoming questions of spatial organisation, i.e. the management of the linkages between the external and internal spatial organisation of cities, regions, nations, and macro-regions such as the European Union. Despite many particularities in world regions such as the emergent system of mega-cities in Asia, the difficult

20 for a more in-depth discussion of the OMC in the context of territorial cohesion policy see: European Spatial Research and Policy (2007) with contributions of: A. Faludi, N. Schäfer, I. v. Homeyer, S. Zirra and J. Buchkremer, J. Farinós Dasí

processes of nation- and region-building in Africa etc., which are quite peculiar to these world regions and not comparable to Europe, many topics discussed in urban and territorial development are becoming similar world-wide. Visions and strategies of urban and territorial development in Asia, for example, focusing on competitiveness and innovation, regional disparities and polarisation, demographic changes and aging, climate change, energy prices and hazards prevention, as well as cultural diversities and assets, are essentially the same as discussed in Europe. As a consequence, it is now the time, or even overdue, to broaden the dialogue and discussion beyond national and world-regional boundaries and to open scientific and political debates to worldwide mutual exchange and learning.[21]

21 cf. Schön (2008)

References

Ahlke B., Görmar W., Hartz A. (2007): Territoriale Agenda der Europäischen Union und transnationale Zusammenarbeit. In: Informationen zur Raumentwicklung IzR, No. 7/8, 449-463.

Battis U., Kersten J. (2008a): Europäische Politik des territorialen Zusammenhalts: Europäischer Rechtsrahmen und nationale Umsetzung. Bonn: BBR.

Battis U., Kersten J. (2008b): Europäische Politik des territorialen Zusammenhalts: Europäischer Rechtsrahmen und nationale Umsetzung. In: Umwelt- und Planungsrecht (UPR), No. 6.

BBR (ed., 2001): Study Programme on European Spatial Planning. Final Report. Forschungen Heft 103.2, Bonn. http://www.bbr.bund.de/cln_007/nn_147036/EN/Publications/Forschungen/backnumbers.html

Böhme K., Schön P. (2006): From Leipzig to Leipzig – Territorial Research Delivers Evidence for the New Territorial Agenda of the European Union. In: disP 165, No. 2, 61-70.

Consolidated versions of the Treaty on European Union and the Treaty on the Functioning of the European Union; Official Journal of the European Union, C 115, Vol. 51, 9 May 2008 http://eur-lex.europa.eu/JOHtml.do?uri=OJ:C:2008:115:SOM:EN:HTML.

Dasi J.F. (2007): Is the OMC a Way to Improve Good Governance in Europe? The OMC as Practice for Territorial Governance. In: European Spatial Research and Policy, No. 1, 85-98.

Davoudi S. (2006): Evidence-Based Planning – Rhetoric and Reality. In: disP 165, No. 2, 14-24.

Delors, J. (2004): Erinnerungen eines Europäers. Berlin. (orig.: Delors, J. (2004): Mémoires. Plon).

Eltges M., Hatzfeld U., Schön K. P., Selke W. (2007): Einführung. In: Informationen zur Raumentwicklung IzR, No. 7/8, I-VIII.

European Commission (1999): ESDP – European Spatial Development Perspective. Luxembourg.

European Commission (2001): European Governance – A White Paper. COM(2001) 428 final.

European Conference of Ministers responsible for Regional Planning (CEMAT) (2000): Guiding Principles for Sustainable Spatial Development of the European Continent. Hannover; reprinted in Informationen zur Raumentwicklung IzR, No. 7, The Council of Europe's Spatial Planning Policy, annex 1-21; http://www.bbr.bund.de/cln_005/nn_21288/EN/Publications/IzR/2003/7Anhang1,templateId=raw,property=publicationFile.pdf/7Anhang1.pdf

European Spatial Research and Policy (2007): No. 1, 11-24.

Faludi A. (2007): Now More Than Ever: The Open Method of Coordination in EU Territorial Cohesion Policy. In: European Spatial Research and Policy, No. 1, 11-24.

Kersten J. (2008): Mindestgewährleistungen im Infrastrukturrecht. In: Informationen zur Raumentwicklung IzR, 1/2, 1-15.
Schäfer N. (2007): The ‚Territorial Agenda for the European Union' and the OMC – Close Relations or No Relations? In: European Spatial Research and Policy, No. 1, 25-41.
Schmidt-Seiwert V., Porsche L. (2007): Kartensammlung zur Europäischen Raumentwicklung – analytische Illustration der Territorialen Agenda der EU. In: Informationen zur Raumentwicklung IzR, No. 7/8, 441-448.
Schön K. P., Selke W. (2007): Territoriale Agenda der EU – ein Ansatz für ein neues Planungs- und Entwicklungsverständnis in Europa. In: Informationen zur Raumentwicklung IzR, No. 7/8, 435-440.
Schön K. P. (2008): Spatial and Urban Development in Asia – Introduction. In: Informationen zur Raumentwicklung IzR, No. 8, I-VI.
Territorial Agenda of the European Union (TAEU) (2007): Towards a more competitive and sustainable Europe of diverse regions. Leipzig, May 2007. http://www.bbr.bund.de/cln_007/nn_169708/EN/ResearchConsultation/SpatialPlanning/SpatialDevelopmentEurope/EuropeanSpatial_20DevelopmentPolicy/01__Publications.html

12 Reflections on Leitbilder

Wendelin Strubelt

In the field of science and planning today, we can observe that most of the internationally used terminology stems from the English "commonwealth" of words. Within scientific and day-to-day phrases, there exists an English dominance, while the power of German as a former *lingua franca* of science - which was never totally true - is gone, forever buried by the megalomania of the Nazis and their misuse of German traditions and language (cf. Klemperer's critique of the *Lingua Tertii Imperii* / LTI).

In addition, the impact of German words on the normal usage of language in the UK and the USA was still minor. Words like *Angst, Ersatz, Hinterland, Weltanschauung, Weltschmerz* or *Zeitgeist* belong to the rare examples of a more or less learned usage of English, while *Kindergarten* in the accustomed form as kindergarden is one of the rare, friendly examples implanted into English colloquial language. From time to time, words like *Blitzkrieg* – sometimes just *Blitz* – or *Panzer* show up as language relics of the Third Reich – especially in English newspapers – causing unpleasant memories both verbally and intentionally.

Quite contrary to this rare or very special presence of German in the English and American language, we all know about the dominant influence of English or American terms on the German language, leading to complaints about this mixture of languages comprising *Denglish*. However, there exists a common tendency of the German language as practised by the Germans to absorb foreign terms. And from a historical perspective in the German context, this has not been a singular development over the last three hundred years. While the influence of Latin and Greek on German as a learned language is quite well known – sometimes even seen as a form of "tyranny" from outside – we tend to forget that during the formation of German as a literary language – fundamentally analysed by the English German scholar Blackall – at the end of the 18th century, the French language also had an overwhelming impact on the German language. An impact which forced many language purists to defend German by creating new German terms – confer the dictionary by Campe – to fight the foreign influence; their nickname was "word-washers". In addition, I should not forget to mention that such movements to defend the own original language against foreign influence are tendencies we can observe in almost any written language and in the context of the many people speaking their "own" language.

In regard to the fact that Germany was one of the leading countries in the field of spatial planning at the beginning of the 20th century and within an international exchange with other similar developments, e.g. in the USA (cf. the Tennessee Valley Authority), it is quite astonishing that in this research and policy field no German *technici termini* have had an international impact, have left a language footprint. Even a well known theory or concept like Christaller's *Zentrale Orte* did not cross over with its original language branding, but was translated into English as "central places". Perhaps not the least because it was easily translatable. And not to forget, the majority of related terms was developed shortly after the NS regime, which did not only isolate the German scientific scene, but afflicted it negatively for a long time to come. Thus, the German planning terminology, the usage of specific terms, developed in a quite isolated way rarely crossing language borders, but nevertheless it was not as much influenced by English terms as other fields. Terms like *Verkehrsberuhigung* (traffic calming) or *Fußgängerzone* (pedestrian zone) are complicated, less apt for a language crossover, and hard to pronounce, not least on account of their bureaucratic origin of phrasing.

However, this is a pity especially for such a "nice" word like *Leitbild/Leitbilder*, because it is quite similar to the term *Leitmotiv*, having been present in the English language for a long time as *leitmotif*. This word is a combination of a German and a French word of Latin origin, combining the lemma of German "leiten", since the times of old historic German meaning to guard or to direct, with the word "motiv", denoting a meaningful topic, a sujet or a trait, a hunch or an element, hardly even a movement in a symphony. It actually stems from a musical background, because the term *Leitmotiv* was first used to circumscribe a musical phrase which is basic or returns from time to time, thus structuring a composition. In dictionaries, the word is connected to Richard Wagner, who is assumed of having created or used it for a characteristic trait of his music, e.g. typifying a character with a musical theme. The musical background of the word was soon transferred to the literary scene with the same intention, namely referring to a contextual or thematic topic guiding a piece of literature. In some way it can be said that this crossover from German into English happened at a time when the acceptance of German scholarship in the fields of musicology and philology was a real and dominant one - worldwide.

Among the numerous German nouns combined with "leiten" such as *Leitfaden* (main connecting thread), *Leitgedanke* (central idea), *Leithammel* (bellwether), *Leitlinie* (guideline), *Leitplanke* (reflector post), *Leitsatz* (basic principle), *Leitspruch* (motto), *Leitstelle* or *Leitstand* (control center), *Leitstern* (lodestar), *Leitstrahl* (control beam), *Leittier* (leader of the pack), *Leitwährung* (reserve currency), *Leitwerk* (tail unit), and *Leitzins* (base rate) with the meaning of directing and closely affiliated to the English lead/leader, *only Leitmotiv* has made its way into the English usage. Nevertheless, the German word *Leitartikel*, the main article or comment in a newspaper called "leader", was directly transferred from English, namely from the "leading article", into German using the connection to many other nouns with the component *Leit...*

And *Leitbild/Leitbilder* – my topic? *Leitbild* is certainly a younger sibling in the family of German words. It is more a word of the 20th century, younger than *Leitmotiv* and much younger than *Leitfaden*, known already at about 1800, as the dictionaries of Adelung and Campe testify, internally taking up or quoting the saga of Theseus and Ariadne, in which the red woolen thread enabled Theseus to escape Minotaurus' labyrinth. *Leitbild* as another picturesque German word was created too late, namely at a time when the international power of German as a leading language was already gone. Thus, we additionally have to look for translations into English when we try to transfer its meaning into the English-speaking hemisphere.

According to my sensitivity for a language its translation into English as guiding principle is too restricted. It focuses on the element of guiding or directing and the meaning of principle, being more connected with the world of essence, truth, or even law while neglecting the much less strict and open meaning of the word "Bild", which means picture and therefore, in a more metaphorical way, could be better translated into concept, *image, motto, orientation, perspective, theme or vision*, instead. And in the German context of planning it rather aims toward a proposed orientation rather than at essential or principal requirements which have to be observed or even followed when spatial planning is in actu. *Leitbilder* as a conceptual scheme try to formulate how spatial development and planning should follow general intentions or orientations without prescribing in detail what should be done in a special place or moment. Thus, concepts/Leitbilder can recur to well known recipes and on new, proposed developments, following new orientations and trying to implement them in the future. In this way, *Leitbilder* can rely on old paradigms (paradigmata) or they can offer new ones. They can be conservative or revolutionary ones (revolutionary breakthroughs) depending on their status in the process of social and political developments over time. They can direct intentions to undiscovered or suppressed orientations and interests. Thus, they are generally full of values as results of already met decisions or prior to decisions to be made in the future. And as I have already said, they are not fixed or even detailed regarding their intentions. They will offer a general framework for planning orientations or decisions to be made rather than offering a clear-cut prescription or even a legal rule that can be relied on – even in court.

I will try to prove this by looking at some *Leitbilder* which we can find in the German planning discussion.

After 1945, one of the dominant paradigms for spatial development and planning was the idea of the *Gleichwertigkeit der Lebensverhältnisse* (equality of living conditions) – another example of important terms from the German spatial planning terminology that are hard to translate – which was deduced by referring to the constitution as a general target for all levels of spatial planning. There were no clear-cut prescriptions, but with the help of statistical measurements or indicators the deviation from the average was analysed and recognised in order to give orientation and advice for political and planning actions in the form of a decisive level. Neither should this level be transgressed, nor should anyone fall behind it. Thus, e.g. in some programs of regional development the fulfilment or non-fulfilment of a given level was the legal ground for distributing money in the form of subsidies. In the beginning of the Federal Republic of Germany, this apparently fair concept was heavily influenced by the conviction that the development of the more rural and in many statistical measures backward regions should be supported while the urban regions, being rich, could and should bear the burden. The difference between *Stadt und Land*, between urban and rural regions was to be equalised mainly by creating or transferring jobs and investing in infrastructure. In some older ways of phrasing such programs there was still an ideological connotation of anti-urbanism by stressing the "unhealthy" quality of urban living conditions especially in big cities *(Großstädte)*, while the social quality of life in rural areas and in the smaller and medium sized cities was considered to be the genuine way of living. From the beginning of the Federal Republic, after 1945, we can even observe that the overall terminology of regional development was ideologically or politically focused on the process of decentralisation, a forerunner of some sort of equalisation policy. In some way, this was a continuation of the anti-urban attitude by which urban and regional planning in the times between the two world wars was characterised all over Europe, e.g. in Germany and in England as well. But in Germany during the Nazi regime it was more or less the dominant ruling doctrine, an outspoken official ideology (*Blut und Boden* / blood and soil), not only a *Leitbild*. However, it is interesting to know that while this ideology was preached to the public it faced the reality of a continuous growth of cities and a continuing migration from rural regions to cities. It has to be mentioned that the military power of the NS regime was not fueled by an agriculturally based society of rurally rooted *Herrenmenschen* (another almost untranslatable German word, sometimes interpreted as "master race"), not even by normal farmers, aside from their being drafted as soldiers, but rather by the heavy industrialised urban areas and the workers living there. But not only in West Germany the aim was to guarantee an equalised settlement structure. In a similar way, regional planning – called territorial planning - in the Soviet zone of Germany (the later German Democratic Republic) was influenced by the overall target of a balancing policy between urban and rural areas, however based on a Marxian background, which can be derived from the Communist Manifesto and from its implementation by Lenin. From the beginning, it was a strategic element of centralist planning in the communist states, as long as they were able to afford it economically. Later it was overruled by economic considerations of separate sectors, neglecting any cross-sectoral considerations, not to mention any extra-sectoral ones.

By reviewing the intentions and the results of such planning directives, not only as orientations, but as clear political intentions based on ideological convictions, in more or less positive terms within the communist ideology, but in more negative terms within the Nazi ideology in regard to its inhuman and racist aspects, we can discover a considerable difference (almost a schism) in both cases. This difference – meaning hiatus or bifurcation – takes place between the outspoken political or ideological intentions and the social and spatial reality.

But in the Federal Republic of Germany after 1945, this was the case in many relations as well. This is reflected by the fact that *Leitbilder* or even *Leitziele* (main targets), as *Gleichwertigkeit*, laid down in laws, in the constitution, and in special laws as well, especially in the *Raumordnungsgesetz* (Federal Regional Planning Act), do not have the character of clear-cut prescriptions. They have the character of so-called *unbestimmte Rechtsbegriffe* (undetermined legal clauses) delivering some sort of legal frame of reference or orientation via *Leitbilder*, while the concrete implementation is open to policy-making on different federal levels, and open to final political decision-making especially at the local level, where spatial policies turn into concrete developments. However, courts might prove whether the actual project will correspond to the intentions of the legislator. In other words, the overall aim to create *Gleichwertigkeit* did not mean that such a development was pursued in a more or less complete way. It was rather an orientation, while the real spatial development favoured by social and especially economic developments on the corporate and individual level could proceed into quite opposite directions. Thus, *Leitbilder* refer to orientations, sometimes only (!) to visions of the future, while the concrete present might be more difficult to explain or to understand than the fulfilment of such orientations, not to speak of visions or prophecies.

If we remember that the idea of *Gleichwertigkeit* was not least posted and enacted as an intention to balance urban–rural disparities, we can observe that over the long period of economic and spatial reconstruction after World War II there was never an actually planned or intended spatial "turn" as designed by the many demands for decentralisation, but rather a continuous process of urbanisation. In the context of this process, the cities grew into city regions or even big agglomeration areas ("suburbanisation") while the rural areas, if not attached to urban or tourist developments, were subject to change caused by the decreasing importance of agriculture as an economic sector. And in addition, urbanism as a way of life became the dominant orientation of people, especially through mass media. The final result was an almost complete loss of difference between urban and rural constellations, resulting in some sort of ubiquity of urbanisation in every corner of the country, at least in the very densely populated Federal Republic of Germany. *Gleichwertigkeit* turned out to be a ubiquitous process of urbanisation leaving the rural side almost completely dependent on cities and the societal and economic developments there. In some way, it was neither a process of centralisation nor of decentralisation, but rather a process of homogenisation by urbanisation, by the penetration of urban ways of life in every corner of the country. This was not at least favoured by the enormous investments in infrastructure, especially in traffic infrastructure, but in social, especially educational infrastructure as well. The idea of the fifties, but especially of the sixties, that everybody should reach the German highway system within at least 30 minutes was more effectual for the homogenisation of the country than any other spatial policy. The target with the term *Gleichwertigkeit* turned out to be fulfilled, but quite differently from what the theorists of decentralisation had had in mind or hoped, namely the reduction of the influence of cities in contrast to a designed and growing importance of rural areas or medium-sized cities – these visions failed. Within the context of the concepts for spatial development, the aim of equalisation was the dominant one until the beginning of the 80s, when the idea of sustainability became more important. While the principle of *Gleichwertigkeit* was dominant in the programme for spatial planning of the 70s, we can observe a concentration on the idea of a crossover policy in the "Programmatische Schwerpunkte der Raumordnung" (Main points for spatial planning) of 1985, connecting spatial policy with environmental policy. In addition, according to the current spatial development, the abstract difference between *Stadt und Land* (urban and rural areas) was differentiated so the old industrialized urban areas and the peripheral rural areas were identified as new problem areas. This differentiation defined new *Problemräume* (problem areas) and further developed the idea of relative *Gleichwertigkeit* by concentrating on those areas really in need. And after the reunification of Germany every endeavour was made to repeat this in the former GDR, the now new German *Länder* (federal states) in order to overcome the spatial differences between East and West. However, in the accompanying Guidelines for Regional Planning (Raumordnungspolitischer Orientierungsrahmen) the aim of *Gleichwertigkeit* was now considered not to be a short-term target, but rather a long-term one, another way of relativisation. Still, the *ländlicher Raum* (rural areas) remains a special spatial category and is supported especially by those agencies feeling responsible for agriculture, but it is not a homogeneous spatial category anymore. Its character rather depends on its vicinity to urban areas or on its character as tourist area rather than on a dominant impact by agriculture. There is a diversity of so-called rural regions. The connotations we normally have had with the term "rural" refer to a societal and economic situation or constellation no longer existing, as we have to face the continuous fading away of agrarian societies in Europe, especially in Western Europe. Therefore, in any case a very detailed analysis of each region is needed in order to understand the paths of the past and the possible paths for future development. The way and the times of simplification via simple one-dimensional concepts are over. And the new *Leitbilder* we speak about today, and in this book for the first time, recognize where the real dynamism of spatial development is located and which help these dynamic regions can offer to other ones, similar to common responsibilities / partnerships. This is more or less a realistic turn in spatial policy by recognizing the reality of societal developments with regard to their spatial incidence rather than aiming at a target which reflects more wishful thinking via spatial images relying on past times, as opposed to a necessary approach for further adjustment concordant with ongoing developments. Maybe it is an exaggeration, but for Germany these new *Leitbilder* in their realistic approach have, for the first time, coherently broken with the old urban-rural schism, which always led to a preponderance of the development of so-called rural areas while not recognizing where the strengths of a country would come from, and how they could be maintained while not neglecting other areas which have to fulfil other functions.

However, in relation to urban development, similar fundamental structural changes can be observed. Apart from the general homogenisation of our society via urbanisation as a societal process and the ubiquity of urbanisation even in former rural or not yet urban regions, we can observe, as already mentioned, a similar differentiation process between cities in Germany depending on their share in booming or declining economic regional structures. In the German context, this differentiation started in the 70s, appearing as the so-called "South-North decline" differentiating the old industrial urban areas of the North from the more modernised industrial urban structures of the South. This resulted in the problem of the future path of urban development no longer being coherent. Since then, processes of growth and of shrinking have existed simultaneously, demanding for new, regionally adapted orientations. This was reflected in both conceptual schemes for spatial policy of the 70s and the 80s.

But what are the *Leitbilder* for urban development, apart from their juxtaposition to the competing rural development?

As already said, the continuing process of suburbanisation favoured by a housing policy supporting private ownership was not only seen as an unwished urbanisation process into former rural areas known as urban sprawl (in German "Siedlungsbrei" or "Zersiedlung"), but as an insult to the existing urban structures as well. However, it is a strange irony that the *Leitbild* derived from these criticised developments presented the *Leitbild* of creating urbanity by density (*Urbanität durch Dichte*). And how was density conceived and operationalised?

By building high-rise apartment blocks at the fringes of the cities! The well-known more than problematic results of these urbanistic innovations – in respect to the architectural shape of cities, but especially in regard to social life – do not have to be discussed in detail here. Because by now it is a common experience that such architectural design may be in line with the specialised needs of some inhabitants, but, as a general concept it is incapable of creating vivid and accepted urban environments. At least in European contexts this is well known, because additional problematic urban areas were "created" with the "help" of this concept.

In its apparent simplicity this idea can remind us of another simple *Leitbild* being prevalent after World War II, namely the *autogerechte Stadt* – the car-friendly city. Both *Leitbilder* follow a simple idea with some sort of face validity, not completely wrong, but not completely fit for a – in its old meaning – sustainable future. In essence, this is the problem of any such *Leitbilder*. As they have to be very abstract in order to include as many opportunities or cases as possible, they mostly do not fit every individual case, they have to be adjusted, requiring inspiration and imagination. However, this is mostly lacking in many cases, as we know from the past. And today, what is the prevalent *Leitbild* or paradigm for urban development? In Germany, we nowadays talk about the *Europäische Stadt*, the European city with its treasure of a lively variation of private and public spheres, with its active citizens and its civic traditions. This model is considered to be a special *Kulturgut*, meaning cultural heritage, which needs to be preserved for the future. As a countermodel to some perceived developments of the American or Asian city, not being homogenous developments in themselves, this European model might be suitable as an overall image concept. However, if we look at the reality of European cities, we find European cities and not the European city. Therefore, in any case, we have to be aware that such general concepts in their emphasized simplicity carry with them apparent and hidden ideological assumptions, sometimes even pure ideologies. We should be suspicious towards such ideologies (in German, we say "Ideologieverdacht"), should evaluate their face validity , and should be careful in regard to their consequences, sometimes not at all unintended ones.

This brings us back to use and function of *Leitbilder*. Within the German context with its prevalent federal structure there is no national duty or authority for issuing regulations commendatory for every level of the Federal Republic. We rather have a "landscape" of fragmented authorities corresponding to each other, but not depending on each other. For the federal level being responsible for the Republic in toto this has led to a culture of issuing recommendations, e.g. in order to guarantee the *Gleichwertigkeit der Lebensverhältnisse* - in addition by materializing these recommendations through thematically oriented subsidies. These policies or this policy-making by persuasion rather than by command functions via general and sometimes even detailed *Leitbilder* setting the scenery. This is the function of *Leitbilder*, and under the premises of this function they have to be taken as valid and sometimes even as powerful tools for structuring the political scenery. As such, *Leitbilder* are an essential element of policy-making in our federal system. They offer orientations, sometimes visions, but they are not prescriptions in the form of guidelines or even legal / administrative recipes. They have to be proven by concrete solutions and policies because they are an offer rather than a blueprint.

Returning to the beginning, in better times the German term *Leitbild* could have been a perfect term fit for a crossover into other languages. For better communication across cultures and borders we now have to retreat to our Latin heritage by using the terms concept or vision (*Konzept* or *Vision*) if we want to transfer the meaning of our terminology into other contexts – not a bad choice at all and not a Latin tyranny.

References

Adelung J. C. (1793 – 1801): Grammatisch – kritisches Wörterbuch der Hochdeutschen Mundart. 4 vols. Leipzig.
Blackall E. A. (1966): Die Entwicklung des Deutschen zur Literatursprache 1700-1775. Stuttgart.
Bundesamt für Bauwesen und Raumordnung (ed., 2006): Informationen zur Raumentwicklung, No. 6/7
Bundesministerium für Raumordnung, Bauwesen und Städtebau (ed., 1985): Programmatische Schwerpunkte der Raumordnung. Schriftenreihe 06 Raumordnung. No. 06.057. Bonn.
Bundesministerium für Raumordnung, Bauwesen und Städtebau (ed., 1993): Raumordnungspolitischer Orientierungsrahmen. Leitbilder für die räumliche Entwicklung der Bundesrepublik Deutschland. Bonn.
Bundesamt für Bauwesen und Raumordnung and Bundesministerium für Verkehr, Bau und Stadtentwicklung (eds., 2006): Perspektiven der Raumentwicklung in Deutschland. Bonn/Berlin.
Bundesamt für Bauwesen und Raumordnung (ed., 2006): Informationen zur Raumentwicklung No. 11/12.
Butler E. M. (1958): The Tyranny of Greece over Germany. Boston.
Campe J. H. (1807 – 1811): Wörterbuch der Deutschen Sprache. 5 vols. Braunschweig.
Campe J. H. (1813): Wörterbuch zur Erklärung und Verdeutschung der unserer Sprache aufgedrungenen fremden Ausdrücke. Ein Ergänzungsband zu Adelung's und Campe's Wörterbüchern. Braunschweig.
Encarta (1999): World English Dictionary. London.
Grimm J., Grimm, W. (1984): Deutsches Wörterbuch. 33 vols., reprint. München.
Häußermann H. (ed., 2000): Großstadt. Soziologische Stichworte. 2nd ed. Opladen.
Klemperer V. (1949): LTI. Notizbuch eines Philologen. Berlin.
Langenscheidts Enzyklopädisches Wörterbuch der englischen und deutschen Sprache (1962). Berlin.
Münkel D. (ed., 2000): Der lange Abschied vom Agrarland. Göttingen.
Pons/Collins (1999): Großwörterbuch für Experten und Universität. Deutsch/Englisch, Englisch/Deutsch. Stuttgart.
Schäfers B., Zapf W. (2001): Handwörterbuch zur Gesellschaft Deutschlands. 2nd ed. Opladen.
Schivelbusch W. (2005): Entfernte Verwandschaft. Faschismus, Nationalsozialismus, New Deal 1933 – 1939. München.
Siebel W. (ed., 2004): Die europäische Stadt. Frankfurt.
Strubelt W. (2008): Städte in Europa - Reflektion eines Sozialwissenschaftlers. In: Deutsche Akademie für Städtebau und Landesplanung (ed.), Metropolregionen. Neue Dimensionen der Europäischen Stadt? Almanach 2007/2008. Berlin, 309 - 326.

Appendix

New Concepts for Spatial Development in Germany

Concepts for spatial development show the desired future state of the spatial structure. They are based on the analysis and evaluation of the current state and from time to time adjusted to the changing social and economic framework conditions. They are an important coordinating instrument of spatial planning. According to the German Federal Regional Planning Act, the Federation and the states are supposed to jointly develop "concepts for the spatial development of the Federal territory" on the level of Federal spatial planning. The last concepts on spatial development had been formulated following the German reunification as "Raumordnungspolitischer Orientierungsrahmen" (ORA, 1993 – Guidelines for Regional Planning) and as "Raumordnungspolitischer Handlungsrahmen" (HARA, 1995 – Political Framework for Regional Planning). They served to improve the spatial planning strategy to harmonise the living conditions in the old and new Federal states.

Now, about one decade after the adoption of ORA and HARA, the framework conditions for spatial planning policy have changed again and advanced. Similar as nearly all political areas, it has to react to basic trends and changes in economy and society as well as to related large reform discussions in Germany as they contribute to considerable changes of spatial structures and uses. This goes for globalisation processes in the same manner as for reunification-related transformation processes and the coping with the demographic change. Furthermore, the European dimension of spatial planning has considerably gained in importance.

The 31st Standing Conference of Federal and State Ministers Responsible for Spatial Planning (Ministerkonferenz für Raumordnung, MKRO) therefore decided on 13 October 2003 "… to check the necessity of advancing the spatial development concepts for the Federal and the European territory". Based on the analysed results and the recommendations of action of the BBR's Spatial Planning Report 2005, the Federal Ministry of Transport, Building and Housing in 2004 launched a large discussion process involving all actors of spatial planning and spatial research in further developing the spatial concepts. The result, which was based on a large consensus, was presented at a conference in September 2005 before being further treated in the committees responsible for the horizontal and vertical coordination of spatial planning. As a result of this large consensus-building process, the new *"Concepts and strategies for the spatial development of Germany"* were accepted by the MKRO on the 30 June 2006 as the joint spatial planning and development policy of the Federation and the states. It also declared its willingness to implement these concepts.

The new concepts for spatial development in Germany are oriented towards the following priorities:

• growth and innovation,
• securing services of public interest,
• conservation of resources, shaping of cultural landscapes.

Concept 1: Growth and innovation

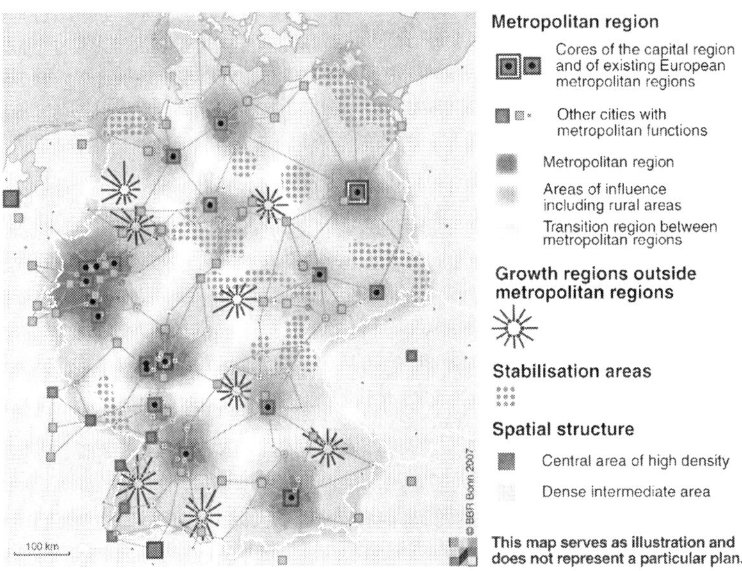

Fig. 1: Growth and innovation. Source: BBR/BMVBS (ed.): Perspectives of Spatial Development in Germany. Bonn/Berlin, November 2006.

By the concept "Growth and innovation" spatial planning subordinates its objectives and strategies to the national objective of promoting overall economic growth. Apart from the classical balancing objective of promoting regions with development weaknesses in order to adjust the living conditions, which will not be abandoned, spatial planning wants to support specific strategies to promote strong regions which are said to be growth motors for the overall economic development. Existing development and innovation potentials are to be strengthened by expanding the infrastructure, fostering certain innovative sectoral and knowledge structures and by fostering education and research etc. In doing so, the locations are not to be considered in an isolated way but the potentials in a region are to be linked up and bundled and growth partnerships are to be founded. At the same time, within a larger regional environment, the growth centres of a region are to assume responsibility for weaker parts in the hinterland and the periphery. They are also to practise solidarity by involving them in economic success and by contributing to stabilising the declines.

The concept ties up to the ORA concept concerning *"European metropolitan regions in Germany"*. Following the ORA and other activities under the Federal action programme "Demonstration Projects of 24 Research News 1/2007 Spatial Development", 11 metropolitan region initiatives have established whose centres are represented in the concept. The centres are the locations with the highest concentration of metropolitan functions. The networking of centres with other locations with important metropolitan functions is exemplarily indicated. Most of them are to be found in metropolitan regions. However, there are also first signs in peripheral places, even in the more structurally weak "stabilisation areas", which are to be integrated in metropolitan growth strategies. Areas of metropolitan influence including rural areas give a rough impression of possible large-area communities of shared responsibilities which cover all types of areas.

"Growth regions outside metropolitan regions" considerably contribute to the overall economic growth. These rather rural areas with important medium-sized cities often represent important regional centres of innovation and specialised locations of technology and are to be supported in sharpening their specific profile and development.

In comparison to this, *"regions with a need for stabilisation"* show a development which is particularly below the average. They are mainly rural areas in a peripheral or border position with insufficient employment opportunities and often antiquated industrial structures. In these regions, regionally adapted strategies are supposed to promote endogenous development potentials in order to prevent a further decline of these areas. Metropolitan regions and growth areas in terms of large regional partnerships have a special responsibility here.

Concept 2: Securing services of public interest

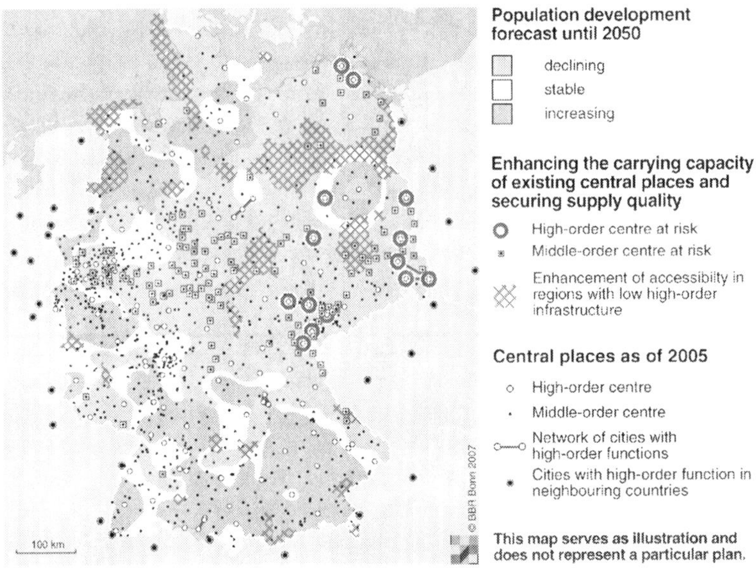

Fig. 2: Securing services of public interest. Source: BBR/BMVBS (ed.): Perspectives of Spatial Development in Germany. Bonn/Berlin, November 2006.

The concept "Securing services of public interest" is the response of spatial planning to the demographic change, which confronts many regions with the problem of ensuring an adequate and well accessible supply of services and infrastructure. Decrease and ageing of the population above all in anyway sparsely populated regions are a big challenge for public households under the current economic framework conditions: the so far good quality of supply has to be maintained. Although the carrying capacity of facilities starts to decrease, the basic supply especially in the health, education and public transport sector has to be guaranteed with minimum standards, which still have to be defined. Through new, temporally and locally flexible forms of organisation and supply, the quality of supply is furthermore to be adapted to modern possibilities and standards and, if possible, to be improved.

In many regions the population decline requires to tighten the *system of central places*. It is supposed to form the basis for an efficient concentration of facilities and services for the public also in the future. The identification of the *carrying capacity of central places is based on the demographic development by 2050*. If the population figures in the service areas of higher-order and middle-order centres, which are defined in Federal state and regional planning, fall below certain thresholds due to a strong population decline (e. g. 300,000 inhabitants in higher-order and 35,000 inhabitants in middle-order service areas), these centres are said to be endangered regarding their carrying capacity. Regions with a high concentration of such at-risk central places will face the most pressure from problems to adjust their central place system. Spatial planning must establish early measures in order to increase service areas in complying with minimum standards of accessibility by combining service areas. It must be noted that even today some regions do not meet the standards for access to a higher-order centre (e. g. 60 minutes by passenger vehicle or 90 minutes by public transport to the nearest higher-order centre). As *regions with a low level of higher - order centre* infrastructure these regions have to be supported by taking measures to improve the transport infrastructure or by shifting certain functions of higher-order o middle-order centres.

Concept 3: Conservation of resources, shaping of cultural landscapes

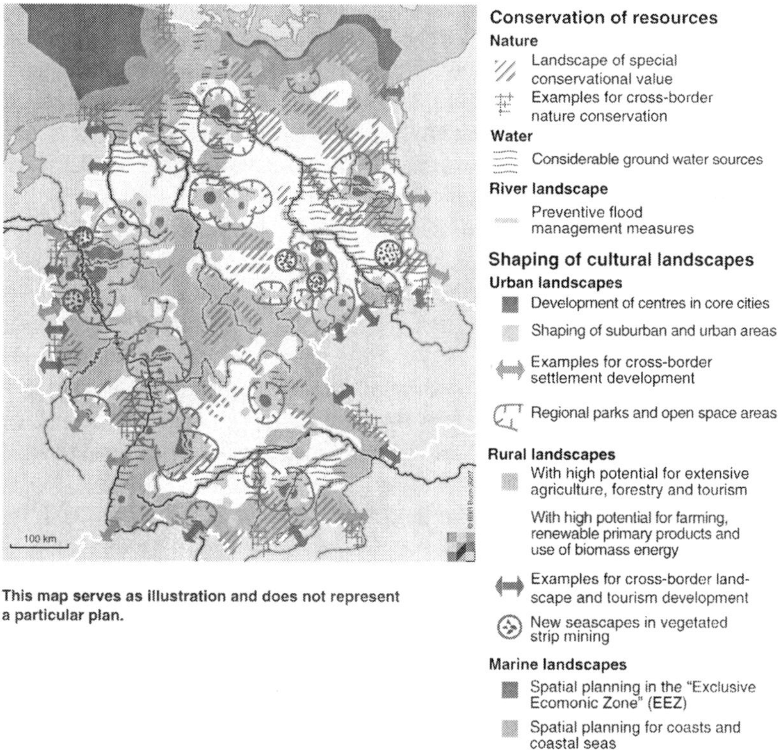

Fig. 3: Conservation of resources, shaping of cultural landscapes. Source: BBR/BMVBS (ed.): Perspectives of Spatial Development in Germany. Bonn/Berlin, November 2006.

The third concept "Conservation of resources, shaping of cultural landscapes" serves to integrate the basic task of spatial planning of caring for a sustainable spatial development into the new concepts. According to this concept, sustainable spatial development will also in the future mean to secure the manifold spatial functions by actively *managing spatial resources* and development potentials in the context of increasing conflicts of use and against the background of the necessity of an economical use with soil as a resource. The task of spatial planning of coordinating the different planning's on an interdisciplinary and supra local level has to be strengthened. The protection of open spaces and the reduction in new demands on land use for settlement and transport is in the foreground here, but also the conservation of nature, water and river landscapes through flood prevention measures.

Similar as in other concepts, spatial planning is supposed here as well to assume more active tasks of shaping and development. The securing and shaping of naturally developed cultural landscapes in line with a sustainable development is therefore a big challenge within this concept. What is needed is the harmonious co-existence of different types of urban, rural and marine landscapes, whereby their ecological, economic, social and cultural functions are permanently preserved and developed.

The *urban landscapes* primarily aim at conserving and developing the functions of centres for city regions. With their relatively high densities and at the same time high attractiveness for close-to-nature living, suburban areas as a subgroup of urban areas have a special development potential which has to be used for quality oriented planning. For densely populated city regions the development of regional parks is highlighted as an example for the shaping task of spatial planning. Developing landscapes near urban centres in a quality oriented way is supposed to provide recreational areas close to settlements.

Rural landscapes with a low population density and a peripheral situation towards centres have their own special development potentials owing to a high close-to-nature landscape attractiveness, which can above all be made usable for tourism. They are at the same time areas in which an extensive, sustainable landscape already today has Concept: Securing Services of Public Interest Concept: Conservation of Resources, Shaping of Cultural Landscapes 26 Research News 1/2007 a relatively high share in the agricultural production and might in the future essentially contribute towards regional development. Due to the high productivity of their soil, other rural landscapes – still – have high potentials for farming, also for newer forms of it like renewable primary products. There are also favourable opportunities to use them for biomass energy. The reclamation of former strip mining areas e.g. affected by brown coal production is an essential instrument of the development of cultural landscapes. In doing so, new lake landscapes are often created for tourist use which may also contribute to the regional economic development.

One task that is of increasing importance is *spatial planning in marine landscapes* since its use for transport, energy and business has increased and has to be harmonised with the objectives of nature and landscape conservation. This is done through "Integrated Coastal Zone Management" (IKZM) by the countries within the 12-mile zone of German coastal waters and through Federal spatial planning as part of the "Exclusive Economic Zone" (EEZ).

What is innovative in the spatial planning policy of the new concepts is a reorientation of the classical tasks of spatial planning:

• The development task is strengthened by activating the specific development potential of all regions, not only that one of the structurally weak but also of the strong regions, to ensure the overall economic growth (Concept 1). The regional balance between strong and weak regions is supposed to increasingly take place on the level of large-area communities of shared responsibility.

• The regional balancing task is reassessed by searching for flexible service standards for the public in order to create equal living conditions without worsening the supply quality but rather to improve it (Concept 2). When making the necessary adjustments, e. g. of the system of central places, minimum standards of accessibility apart from the economic carrying capacity of facilities also have to be taken into account.

• The planning task is underlined by strengthening the competence of judgement. Not only land protection has to be in the foreground but also an active management of resources e.g. to shape the diversity of cultural landscapes (Concept 3). Protection concerns and development potentials have to be equally treated.

What is also new is that the tasks of spatial planning are not explicitly limited anymore to types of areas such as rural areas or agglomerations. They will now also be based on current economic and societal issues which concern all spatial categories in different ways. The concepts in so far always refer to all regions with differentiations:

• Concept 1: metropolitan, growth and stabilisation areas,
• Concept 2: areas showing a population decline,
• Concept 3: urban, rural and marine landscapes.

The new spatial development concepts have been adopted based on a consensus between the Federation and all Federal states. In more detail they provide enough scope for active planning to the actors of spatial planning at all levels and in all spatially relevant planning sectors. The further development will show whether the Federation and the Federal states will implement these framework agreements in the practice of spatial planning.[1]

1 This summary was written by Dr. Horst Lutter, Bundesamt für Bauwesen und Raumordnung (Federal Office for Building and Regional Planning), Bonn, Germany.

Appendix

List of Authors

Jerzy Bański

Born in 1960 in Wolbrom. Full Professor of the Institute of Geography and Spatial Organization, Polish Academy of Sciences (IGSO PAS), and Professor of Maria Curie-Sklodowska University in Lublin, Poland. He received his Ph.D (1996) and his postdoctoral lecture qualification (1999) at the IGSO PAS. His main research interests include: land use, regional policy and socio-economic development of rural areas, changing spatial economic patterns. In 2000 he has become the head of the Rural Areas Study Group at the IGSO PAS and editor of "Rural Studies". From 2006, he has been a president of the Polish Geographical Society and vice-chair of the Commission on Local Development, International Geographical Union. E-Mail: bansi@twarda.pan.pl

Annedore Bergfeld

Born in 1959. 1974 to 1977: studies at the Technical University of Leipzig, 1984: doctorate, 1990: habilitation, 1984-1992: lecturer at the Technical University, 1992-2007: Managing Director of the Institute for Economic and Regional Development Leipzig, since 2007: researcher at the Leibniz Institute of Regional Geography, Leipzig. E-Mail: a_bergfeld@ifl-leipzig.de

Marek Degórski

Born in 1956, MsC (1979). PhD (1988) and postdoctoral lecture qualification (2002) in physical geography. Expert in landscape analysis and environmental planning and management with long experience in the assessment of environmental and human relationships. Conducted research in Poland, Germany, Finland, Norway, Sweden, Estonia, Latvia, Lithuania, Belarus, India and USA. Fulbright Scholar at the State University of Washington, Seattle, and State University of Connecticut, Storrs, in 1991 as well as Visiting Professor at the Harvard Forest, Massachusetts, Northern Arizona University, Flagstaff, Hawaii State University, Honolulu, and Michigan Technical University, Houghton (USA), Ghent University (Belgium), Oulu University (Finland), Roma University (Italy), Swiss Forest Reserch Institute, Birmensdorf, and others. Associate Professor, since 2003 head of the Department of Geoecology and Climatology as well as deputy director at the Institute of Geography and Spatial Organisation, Polish Academy of Sciences. Member of the Polish National Committee for ICSU-SCOPE and UNESCO-MAB, Committee of Geographical Sciences, as well as the Committee of Polish Academy of Sciences for the International Year of Planet Earth. Member of the Executive Committee of IYPE in the International Geographical Union, European Geosciences Union (EGU) and International Union of Soil Sciences (IUSS). Representative of Poland in GEO Working Group of the European Commission (2004-2005). Performer and member of advisory boards and task forces for numerous European projects and programs, e.g. Landscape Tomorrow, Integrated Environmental Monitoring of Europe, Vistula ECONET Development and Implementation, as well as North European Trade Axis. Recently working on numerous projects concerning the 6th Framework Program of the European Commission: PLUREL (Peri-urban Land Use Relationships – Strategies and Sustainability Assessment Tools for Urban-Rural Linkages) and e-LUP (Simulating land-use processes – an interactive e-tool for SIA). E-Mail: m.degor@twarda.pan.pl

Clemens Deilmann

First degree in Architecture at RWTH Aachen (1979), post-graduate diploma at Architectural Association School of Architecture in London (1980), 1981 – 1984: Architect in Medellin, Colombia. 1984 – 1992: freelance Architect with specialisation in ecological design, additionally research assistant at Technical University Darmstadt 1984 – 1989. Since 1992: Head of the dept. Housing and Sustainable Construction at IOER. Since 2006: Head of Research Area Resource Efficiency of Settlement Structures. Research interests: Implementation of Building Ecology into everyday building practice, environmental assessment, analyses of material- and energy-flows in the building sector, urban-structural-type-analyses and development scenarios. E-Mail: c.deilmann@ioer.de

Hans-Jörg Domhardt

Born in 1954. Studies in spatial planning at Dortmund University. 1981-1987: researcher at the Chair for Regional Planning at Dortmund University, Ph.D. studies with Prof. Dr. A. Bloch. 1987-2000: researcher at the Chair for Regional Planning at Kaiserslautern University (Prof. Dr. Hans Kistenmacher). Since 2000: researcher and Academic Director at the Chair for Regional Development and Spatial Planning (Prof. Dr. Gabi Troeger-Weiß). For a number of years his research focuses on spatial-structural and spatial-economic issues. He possesses comprehensive experience in research coordination and project management within numerous international research projects. He has recently contributed to, among others, the subject of basic supply for regions with low-density settlement structure. He is member of the German Society for Demographics and Vice Director of the Steering Group of the Work Group of Hessen/Rheinland-Pfalz/Saarland of the Academy for Spatial Planning and Research. E-Mail: domhardt@rhrk.uni-kl.de

Ludger Gailing

Born in 1976, studied spatial planning at the Universities of Dortmund and Grenoble and graduated in 2002. Since 2003: scientific researcher at the Leibniz-Institute for Regional Development and Structural Planning in Erkner near Berlin. Member of the working groups of research network "4R" (four spatial research institutions in Germany) "Cultural landscapes – Substantialisation for land use planning" and "Cultural landscapes – Suburban areas as cultural landscapes". His research interests are regional governance arrangements of landscape and open space policy as well as institutional change in urban and regional planning. His work in the IRS concentrates on the application of theoretical principles underlying institutions, goods, and paths when analysing the common good cultural landscape, and the examination of institutional case studies of regional landscape policy. E-Mail: gailing@irs-net.de

Günter Herfert

Born in 1948. 1966 to 1970: studies at the Pedagogical University Potsdam; 1970: degree in teaching geography and sports; 1981: Ph.D. in human geography at the Institute of Geography, Pedagogical University Potsdam; 1981 to 1991: postdoctoral research fellow at the Institute of Geography and Geoecology Leipzig. Since 1992: senior researcher at the Leibniz-Institute of Regional Geography, Leipzig. E-Mail: G_Herfert@ifl-leipzig.de

Rupert Kawka

Born in 1969 in Freiburg/Brsg. He studied at the University of Heidelberg and holds diplomas in geography as well as public economy. He received his Ph.D. in geography with a thesis on African towns. 2001 to 2004: contributor at the Halle Institute for Economic Research (IWH). Since 2004: project manager at the Federal Office for Building and Regional Planning (BBR). He is also visiting lecturer at the University of Jena. Some research projects of the past years include new guiding principles for spatial development in Germany, regional costs of living, metropolitan functions in Europe, and the demonstration project "Supraregional Partnership". E-Mail: Rupert.Kawka@BBR.Bund.de

Heiderose Kilper

Born in 1953, Dr. phil., habilitated in Political Science, Director of the Leibniz Institute for Regional Development and Structural Planning in Erkner near Berlin and Professor at the Brandenburg Technical University (BTU) in Cottbus. She has the Chair for Urban and Regional Development in the Faculty of Architecture, Civil Engineering and Urban Planning at the Technical University of Cottbus. From 2002 to 2005: director of an extra-universitarian research institute in Hanover, Chair in the Department of Landscape Planning and Nature Conservation at University of Hanover. From 1990 to 2002: Senior Researcher and Scientific Manager at the Institute of Work and Technology in the Science Center North Rhine-Westphalia and Assistant Professor at the Faculty of Political and Administrative Science/ University of Constance and of Social Science/Ruhr-University of Bochum. Between 1984 and 1990: contributor in various capacities for the government of North Rhine-Westphalia. Her research fields are governance/European multi-level-systems; theories of planning and regulating; structural and developmental policies for regions; demographic change; federalism in the German Federal Republic. E-Mail: kilper@irs-net.de

Tomasz Komornicki

Associate professor at the Institute of Geography and Spatial Organisation at the Polish Academy of Sciences, head of the Department of Spatial Organization and Regional Studies. 1996: employed in the Institute fur Länderkunde, Leipzig, Germany, as researcher. 2005-present: professor at Lublin University (Economic Geography Department). Main topics of interest: transport geography, transport policy, spatial planning and development of border regions; participated in a number of Polish and European research projects including six ESPON projects, IASON and STELLA; deputy president of Polish Geographical Society; member of the Committee on Spatial Economy and Regional Planning, Polish Academy of Science; member of the consulting board for the preparation of the new Concept of Spatial Development of Poland 2008-2033 (in the Ministry of Regional Development). E-Mail: t.komorn@twarda.pan.pl

Thomas Kuder

Born in 1959 in Heilbronn. He studied Urban and Regional Planning at the Technical University of Berlin. After completing his diploma he worked in the Department of Urban Planning in Berlin (Tiergarten) for several years. In 1994 he became a research assistant at the Technical University of Berlin. His dissertation, completed in 2002, addressed the issues of overall concepts (Leitbilder) in Urban Design and Urban Planning. Since 2003: senior researcher at the Leibniz-Institute for Regional Development and Structural Planning (IRS) in Erkner, near Berlin, Department 4: Regeneration of Shrinking Cities. His research activities centre on the issues of orientation and governance in transformation contexts and urban development policies. E-Mail: Kuder@irs-net.de

Sebastian Lentz

Born in 1957. 1976 to 1984: studies at the Universities of Heidelberg and Mannheim. 1984: teacher's exam for Geography and German Philology. 1988: Ph.D. in Human Geography, Institute of Geography, University of Mannheim. 1988 to 2000: lecturer and assistant professor at the Institute of Geography, University of Mannheim. Since 1993: occasional guest-lecturer at Moscow State University, faculty of Geography. 1999: second doctoral thesis on "Segregation through Privatisation of the Post-socialist Housing Market in Moscow." 2001: Professor of Anthropogeography, University of Erfurt. Since 2003: Professor of Regional Geography at the University of Leipzig and Director of the Leibniz Institute for Regional Geography, Leipzig. Member of Sächsische Akademie der Wissenschaften zu Leipzig; Kommission für Landeskunde der Sächsischen Akademie der Wissenschaften zu Leipzig; Corresponding member of the Akademie für Raumforschung und Landesplanung (ARL), Hannover; Member of the Länder-Working Group Sachsen, Sachsen-Anhalt, Thüringen of the Akademie für Raumforschung und Landesplanung (ARL); Member of the Deutsche Akademie für Landeskunde (DAL); Member of the Verband der Geographen an Deutschen Hochschulen (VGDH) – advisor to the executive board since 2004; Member of the Committee for Geography of the German Research Foundation (since 2008). E-Mail: s_lentz@ifl-leipzig.de

Gerd Lintz

Born in 1962. From 1982 to 1989: studies in economics, business administration and sociology at the University of Trier. 1989 to 1994: assistant to the Chair of Urban and Regional Economics at the University of Trier. Since 1995: research associate and senior researcher with the Leibniz Institute of Ecological and Regional Development. Main research interests: rehabilitation and development of old industrial regions, coordination of environmental policy, regional economic policy and spatial planning, environmental policy integration, environmental regional governance, local environmental management approaches. E-Mail: g.lintz@ioer.de

Christoph Scheck

Born in 1981 in Leonberg/Württ. Studies in spatial and environmental planning at the Technical University of Kaiserslautern. His Diploma Thesis deals with the subject of regional cooperation in the Rhine-Main-Area. Since 2006: researcher at the Development Agency Rheinland-Pfalz e.V. in Kaiserslautern and the Chair for Regional Development and Spatial Planning (Prof. Dr. Gabi Troeger-Weiß). His field of work focuses on contemporary issues of regional planning and development, especially the further detailing of spatial development on regional levels (regional land use plan, provision of livelihood). Beyond that, his teaching curriculum includes structural and process analyses in spatial planning. He is currently conducting doctorate research on issues of how to implement large-scale regional cooperation in particular regard to the relationship between urban and rural space. E-Mail: scheck@rhrk.uni-kl.de

Karl Peter Schön

Born in 1952. Since 1992: Head of the European Spatial and Urban Development unit within the German Federal Office for Building and Regional Planning (Bundesamt für Bauwesen und Raumordnung (BBR), Bonn). He studied sociology, economics, and mathematics at the universities of Bochum and Bielefeld and holds a PhD in Sociology from the University of Bielefeld. Before joining the BBR he had worked at the universities of Osnabrück, Berlin and Bielefeld. For the last fifteen years he has focussed on European spatial development building bridges between scientific analysis and policy formulation. From the beginning, he was an active part of the political and scientific cooperation process that led to the basic documents of the Leipzig Ministerial Meeting of 1994, which included the proposal of the German Presidency to build up a European research network on spatial development issues, was part of the drafting groups preparing the European Spatial Development Perspective (ESDP, 1999) and the Territorial Agenda for the European Union (TA-EU, 2007) and its experts' background document on the Territorial State and Perspectives of the EU (TSP) as well as the scientific networking that resulted in the present European Spatial Planning Observation Network (ESPON).
E-Mail: Peter.Schoen@BBR.Bund.de

Przemysław Śleszyński

Born in 1973. 1991-1996: studies at Warsaw University, Faculty of Geography and Regional Studies and Master degree in physical geography. Since 1997: employed in the Polish Academy of Sciences, Institute of Geography and Spatial Organization (IGSO PAS). 2003: received Ph.D. on the socio-economic geography in the same Institute. Thesis: socio-economic transformation of the western part of Warsaw's center after 1989. 2007: postdoctoral lecture qualification published about the economic control function on Poland's territory. Since 1995: member of the Polish Geographical Society (1996-2003 Treasurer of the Main Board). Since 2002: Society of Polish Town Planners. 2002–2003: scholar of the Foundation For Polish Science (FNP). Main scientific research concerning urban, population, electoral and managerial geography (Transformation of Polish cities and regions around 1989, migration and settlement system, economic and social changes, electoral studies and forecasting), also spatial economy, local and regional planning, landscape aesthetics, cartography, GIS. Expert and consultant of government and local administrative institutions (spatial planning, urban economy, cartographical analyses) and also private consulting firms (location and spatial organization of economic studies).
E-Mail: psleszyn@twarda.pan.pl

Wendelin Strubelt

Born in 1943, 1964 – 1966: studies in protestant theology, political science, history, and German literature at the universities of Erlangen-Nürnberg and Tübingen,

1966/67: studies at the University of Colorado, Boulder, 1967-1969: studies at the University of Konstanz, 1969: Graduation M.A., (political science and German literature), 1970 – 1972: research fellow in the Department of Political Science, University of Konstanz, 1973 – 1976: lecturer in social science at University of Bremen, 1976: Graduation (Ph.D., political science and sociology), 1977: assistant professor, Department of Social Science, University of Bremen, 1979: associate professor University of Bremen. Since 1981: Director and Professor of the Federal Research Institute for Regional Geography and Regional Planning. Since 1998: Vice-President and Professor of the Federal Office for Building and Regional Planning. Member of the Academy for Spatial Research and Planning (Akademie für Raumforschung und Landesplanung - ARL) and German Academy for Urban and Regional Spatial Planning (Deutsche Akademie für Städtebau und Landesplanung – DASL). E-Mail: Wendelin.Strubelt@BBR.Bund.de

Gabi Troeger-Weiß

Born in 1958 in Hof/Saale and studied geography and spatial planning at Bayreuth University, the Hochschule für Wirtschaft- und Sozialwissenschaften in St. Gallen, and the Eidgenössisch-Technische Hochschule in Zurich. She holds a Ph.D. in economic geography and regional planning. 1990: Vice Department Manager for different departments within the Bavarian State Ministry for regional planning and ecological issues. Manager of Euregio Egrensis and Forum Zukunft Oberfranken. 1998: habilitation on regional management as a new instrument of spatial planning. 2000: Professor, Chair of Regional Development and Spatial Planning at the Technical University Kaiserslautern. 2003: Visiting Professorship at Klagenfurt University, thematic focus on "Regional Management in the European Context". At her Chair, numerous EU-projects connected to spatial and regional sciences are conducted and a study on growth regions beyond metropolitan areas has recently been completed. E-Mail: troegerw@rhrk.uni-kl.de

Grzegorz Węcławowicz

Born in 1943. 1961-1966: studies at Warsaw University, Master degree in physical geography. 1967-1970: lecturer at the Department of Economic Geography at Warsaw University. 1974: Ph.D. in Economic Geography at the Polish Academy of Sciences, Institute of Geography and Spatial Organization, in Warsaw. 1974-1989: Assistant Professor, 1989-1998: Associate Professor, Since 1998: Professor at the Institute of Geography and Spatial Organization at the Polish Academy of Sciences. After studies at the London School of Economics and Political Studies in 1976-1977, he has delivered occasional lectures or seminars at the Universities of Edinburgh, Glasgow, Manchester, Salford, King's College, and Queen Mary College in London. Since 1991, his research has been supported by different foreign universities: Austrian Academy of Sciences (Institute für Stadt- und Regionalforschung) – 1991, Paris University X-Nanterre – 1992, Universite Libre de Bruxelles – 1993/1994. Since September 1995, he is lecturer at the School of Geography at the University of Leeds. Since the beginning of his work at the Polish Academy of Sciences, Institute of Geography and Spatial Organization, in Warsaw, his field of research has been in the domain of urban geography, social geography, and regional policy. His PhD resulted in a book publication titled "The structure of socio-economic space of Warsaw in 1931 and 1970". His further research on urban areas concentrated on housing issues, the elderly population, and the socio-spatial structure of Polish cities. His current research interests concentrate on two subjects. The first concerns the social and spatial consequences of transformation in Central Europe. Since 1994: Participation in major international research programs. Currently, he is head of the Department of Urban and Population Studies and member of the Regional Science Association (since 1974), the Polish Geographical Association (since 1963, 1997-200 vice-president), and expert and consultant of the central regional administration. E-Mail: g.wecla@twarda.pan.pl

Marek Więckowski

Born in 1971 in Warsaw. 1994-2000: employed at Warsaw University, Faculty of Geography and Regional Studies. 2000: received Ph.D. on regional geography at the same Faculty. Since 2000: employed at the Institute of Geography and Spatial Organisation at the Polish Academy of Sciences. 2002-2004: received the prestigious post-doc position at "Ecole normale supérieure Lettres et sciences humaines (ENS–LSH)" in Lyon founded by the French Ministry of Sciences. 1995-present: member of the Polish Geographical Society (1995-2003 chairman of the Academic Division). Expert and consultant of government and local administration institutions and private consulting firms (spatial planning and organization, tourism development, cross-border development, geographical education). Participated in numerous Polish and European research projects. Main topics of scientific interest: geography of tourism, transport geography, spatial planning and development of border regions, cross-border cooperation, nature protection. E-Mail: marekw@twarda.pan.pl

Peter Wirth

Born in 1959. From 1980 to 1984: studies in geography and mathematics (for high school teaching certificate) at the Teachers Training College in Dresden. 1988: doctorate on questions of land use planning in Dresden. From 1988 to 1991: abstractor and group manager in environmental administration. Since 1992: senior researcher with the Leibniz Institute of Ecological and Regional Development Dresden. His research focuses on regional cooperation, regional governance, non-statutory tools of regional development, management of regional restructuring processes in Europe, ecological renewal of regions characterised by small and medium sized towns. E-Mail: P.Wirth@ioer.de

Horst Zimmermann

Professor Emeritus of Public Finance at Philipps-University, Marburg/Germany. He studied in Cologne, Munich, and at Northwestern University (Evanston, Ill./ USA). Expertise: He has published, among other subjects on public finance in general, on fiscal federalism, esp. local finance, and on the relations between the public finances and the regional structure. Activities: In the Academy for Regional Research and Regional Planning (ARL), he served as Vice-President for four years and as President in 2003-2004. He is member of the Scientific Advisory Board to the Federal Ministry of Finance and serves as Senior Expert in a 4-year-project on administrative-territorial reform in Ukraine for the German Technical Assistance Agency GTZ. E-Mail: horst.zimmermann@debitel.net

Appendix

Portrait of the Editors' Institutions

Academy for Spatial Research and Planning (ARL)

The ARL is concerned with research about spatial structures and developments that are important for Germany, and of international relevance. The ARL is a service institution for basic and application-oriented research. At the same time, it is a neutral forum which promotes discourse about spatial science issues. The ARL network currently consists of approximately 1,000 experts. Their tasks are, among other things, to initiate research and to perform joint research within this network, as well as to encourage a transfer of knowledge.

Academy for Spatial Research and Planning (ARL)
Hohenzollernstr. 11
30161 Hanover

E-Mail: ARL@ARL-net.de
Internet: www.ARL-net.de
Secretary General: Prof. Dr-Ing. Dietmar Scholich

Leibniz Institute for Regional Geography (IfL)

The Leibniz Institute for Regional Geography (IfL) conducts basic research into the regional geography of Germany and Europe and provides – as one part of its educational task – geographical information about spatial structures and their development. Additional research topics are the analysis of spatial structures and development on different scales, as well as the presentation of the geography of Germany abroad.

Leibniz Institute for Regional Geography (IfL)
Schongauerstr. 9
04329 Leipzig

E-Mail: info@ifl-leipzig.de
Internet: www.ifl-leipzig.de
Director: Prof. Dr. Sebastian Lentz

Leibniz Institute of Ecological and Regional Development (IOER)

The Leibniz Institute of Ecological and Regional Development (IOER) carries out interdisciplinary research into the complex interaction of the natural, built and, social environments. Key research areas are the environmental aspects of regional development on a European level, strategies for the ecological restructuring of towns and regions, the environmental impact of land use changes, as well as the material flows of building and housing.

Leibniz Institute of Ecological and Regional Development (IOER)
Weberplatz 1
01217 Dresden

E-Mail: info@ioer.de
Internet: www.ioer.de
Director: Prof. Dr. Bernhard Müller

Leibniz Institute for Regional Development and Structural Planning (IRS)

The Leibniz Institute for Regional Development and Structural Planning (IRS) in Erkner (near Berlin) examines the spatial, social, and economic basics and planning perspectives for regions and towns. Research conducted by the IRS deals with the following topics: strategies for regionalisation, institutional change, regional structures and the restructuring of towns, as well as knowledge milieus in the context of transformation processes.

Leibniz Institute for Regional Development and Structural Planning (IRS)
Flakenstr. 28-31
15537 Erkner

E-Mail: regional@irs-net.de
Internet: www.irs-net.de
Director: Prof. Dr. Heiderose Kilper

Federal Office for Building and Regional Planning (BBR)

The Federal Office for Building and Regional Planning is a superior federal authority in the operational division of the Federal Ministry of Transport, Building and Urban Affairs (BMVBS). On the one hand, the BBR supports the federal government, with sectoral scientific consultation in the policy areas of spatial planning, urban development, housing and building. On the other hand, it is responsible for the supervision of Germany's most important federal buildings at home and abroad.

Federal Office for Building and Regional Planning (BBR)
Deichmanns Aue 31-37
53179 Bonn

Fasanenstr. 87
10623 Berlin

E-Mail: zentrale@bbr.bund.de
Internet: www.bbr.bund.de
President: Dipl.-Ing. Florian Mausbach
Vice President: Prof. Dr. Wendelin Strubelt

Printed in the United States
149935LV00003B/31/P

9 783540 888383